"*Balanced and Barefoot* offers new ways to see, solve, and prevent the reactive behaviors and emotions that cause kids to struggle and parents to worry. Full of practical ways to give kids the kinds of experiences that will help them thrive, this book is a must-read for parents and teachers alike."

—**Tina Payne Bryson, PhD**, coauthor of the *New York Times* bestsellers *The Whole-Brain Child* and *No-Drama Discipline*

"Angela Hanscom explains—beautifully and convincingly—why unrestricted outdoor play is essential to children's healthy sensory, motor, social, and intellectual development, and she shows how we can enable such play in today's world. I recommend this book highly to all parents, educators, and pediatric health professionals, and to anyone involved in making decisions that affect children's lives."

—**Peter Gray**, research professor of psychology at Boston College, and author of *Free to Learn*

"In *Balanced and Barefoot*, Angela Hanscom gives adults permission to release the reigns without fear. Her expertise as an occupational therapist affirms what so many of us know intuitively: children thrive when they have access to rich, self-directed play opportunities."

—**Erin Davis**, director of *The Land: An Adventure Play Documentary*

"Here I am, the 'Free-Range Lady,' and I, too, was often shocked and saddened to see my kids (and others) not know how to organize their own games outside with their friends. This is the book I needed when they were younger!"

—**Lenore Skenazy**, founder of the book, blog, and movement, *Free-Range Kids*

"I am a fan and proponent of Angela Hanscom's nature-based philosophy. Her new book is an eye-opener as she presents from experience, observation, and scientific research not only the benefits of physical activity and free play for children, but also the mental and emotional necessity. In a culture that markets devices that restrict infant movement, endorses early education for toddlers, and over-schedules structured activities for children of all ages, Hanscom's book is both illuminating and timely. Eloquent, logical, and reasoned, *Balanced and Barefoot* is a gift and an important read for educators, caregivers, and parents."

—**Janet Lansbury**, author of *Elevating Child Care* and *No Bad Kids*

"Exposure to nature in early childhood provides a wide range of sensory and psychological advantages for child development. Stimulation of all senses improves physical and mental balance. Natural, free play—as proposed by Angela Hanscom—is essential for nurturing stronger and healthier bodies and minds. I recommend this important book to all parents and teachers, because we need to have the TimberNook experience everywhere."

—**John M. Tew, Jr., MD**, professor of neurosurgery, surgery, and radiology at the University of Cincinnati, and executive director of community affairs at UC Health and UC College of Medicine

"*Balanced and Barefoot* offers a refreshingly straightforward approach that counters the pressures many well-intentioned parents feel in raising children today. It's about backing off and giving children the space to do what they naturally do—to explore and figure out the world, to make decisions, and use their imagination. Being outdoors allows children to learn about themselves, gain confidence and flexibility, learn to problem solve, and get along with others. These are all traits they need for healthy development. I recommend this book for every parent looking to raise an independent, caring, resilient and confident child."

—**Tovah P. Klein**, PhD, director of the Barnard College Center for Toddler Development, and author of *How Toddlers Thrive*

BALANCED
and
BAREFOOT

How Unrestricted Outdoor Play
Makes for Strong, Confident,
and Capable Children

ANGELA J. HANSCOM

New Harbinger Publications, Inc.

Publisher's Note

This publication is designed to provide accurate and authoritative information in regard to the subject matter covered. It is sold with the understanding that the publisher is not engaged in rendering psychological, financial, legal, or other professional services. If expert assistance or counseling is needed, the services of a competent professional should be sought.

Distributed in Canada by Raincoast Books

Copyright © 2016 by Angela J. Hanscom
New Harbinger Publications, Inc.
5674 Shattuck Avenue
Oakland, CA 94609
www.newharbinger.com

Cover design by Debbie Berne; Cover photo and author photo by Millissa Gass at MadLiv'n Design and Photography; Acquired by Melissa Kirk; Edited by James Lainsbury; Indexed by James Minkin

Library of Congress Cataloging-in-Publication Data

Names: Hanscom, Angela J., author.
Title: Balanced and barefoot : how unrestricted outdoor play makes for strong,
    confident, and capable children / Angela J. Hanscom ; foreword by Richard Louv.
Description: Oakland, CA : New Harbinger Publications, [2016] | Includes
    bibliographical references.
Identifiers: LCCN 2015047849 (print) | LCCN 2016005040 (ebook) | ISBN
    9781626253735 (paperback) | ISBN 9781626253742 (pdf e-book) | ISBN
    9781626253759 (epub) | ISBN 9781626253742 (PDF e-book) | ISBN 9781626253759
    (ePub)
Subjects: LCSH: Play. | Outdoor recreation. | Nature. | Child development. | BISAC:
    FAMILY & RELATIONSHIPS / Parenting / General. | NATURE / General. |
    SCIENCE / Life Sciences / Neuroscience. | FAMILY & RELATIONSHIPS /
    Children with Special Needs. | MEDICAL / Allied Health Services / Occupational
    Therapy.
Classification: LCC HQ782 .H346 2016 (print) | LCC HQ782 (ebook) | DDC
    306.4/81--dc23
LC record available at http://lccn.loc.gov/2015047849

Printed in the United States of America

21   20   19

10   9   8   7   6

This book is dedicated to Joelle, Charlotte, and Noah Hanscom.

Contents

Foreword

In recent years, a new nature movement has emerged that includes traditional conservation and sustainability, but gives special attention to the right of every child to the benefits that nature brings to children's physical and mental health and their ability to learn and create. This movement is based on a growing body of scientific evidence—most of it correlative, because research in this arena is relatively new (and overdue), but it all points in the same direction.

As a result, families are joining other families to get their kids outdoors. Regional campaigns have emerged in cities and states in North America and overseas, and many of the nation's mayors are taking action. In education, teachers are creating school gardens and championing nature-based approaches. Mental health professionals are weaving nature into their practices. Within the health care community, a growing number of pediatricians have begun to prescribe nature time to the families they serve.

In *Balanced and Barefoot*, Angela Hanscom, a pediatric occupational therapist, makes a passionate case that nature play is necessary for a truly balanced childhood. Correctly, she does not claim that nature play is a panacea, or necessarily a replacement for other therapies, but that it can be a strong component in prevention and therapy. And for some children, it can make all the difference. As Angela eloquently illustrates, too many of today's children miss out on the full sensory richness offered beyond the walls of a classroom or home. Manageable risk and independent, imaginative play are essential not only to physical health but to the development of self-directed young minds.

For many families and teachers, getting children outside and active is not as simple as it might seem. Fear of strangers (in some neighborhoods fully justified, in other neighborhoods not so much), along with poor urban design, inaccessible parks, and the dominance of electronics—all of these barriers and more are very real. But this book offers rich advice on how parents and professionals can overcome them—in health, education, and the creation of built environments. Balanced and Barefoot will be immensely useful to parents and teachers, pediatricians and pediatric occupational therapists, architects and play space designers, and many others.

Angela Hanscom is a powerful voice for balance.

—Richard Louv, author of *Last Child in the Woods*, *The Nature Principle*, and *Vitamin N*

Introduction

I'm starting a fire in the woods, while happy sounds of children echo through the forest. They are immersed in free play. Well . . . most of them. A young girl approaches me as I start lining up biscuits for the children to roast when they get hungry. "I'm bored," she complains. I look around the woods and see groups of children catching frogs in the marsh, others are working as a team to create a fort using sticks and transparent blankets, and still others are huddled in the corner, captivated by some sort of play scheme. "What's next on the agenda?" she asks. My eyes drift back to the little girl in front of me. She is six years old and stands firm with her hands on her hips. "It is time for free play," I say, without giving it much thought. "There's no schedule."

"What?" the little girl glares at me. "My mom is paying you good money to entertain me!"

At TimberNook, a nature-based developmental program for children, we are not in the business of entertaining children. Quite the contrary. We are internationally recognized for our unique programming and therapeutic qualities, all based on allowing children to play freely in the outdoors with very limited adult interaction. I founded TimberNook four years ago as a way to address many of the sensory issues I was seeing in my youngest clients. As a pediatric occupational therapist, an increasing number of children with problems atypical for their young ages were being referred to me. Both the quantity and the scope of their ailments, from not tolerating wind in their faces to poor balance and uncoordinated bodies to crying and getting easily upset in new situations, alarmed me.

My professional training had taught me that movement—and a lot of it—is key to preventing many of these problems. And through my own research, and in my personal life, I discovered that movement through *active free play*—particularly in the *outdoors*—is absolutely the most beneficial gift we as parents, teachers, and caregivers can bestow on our children to ensure healthy bodies, creative minds, academic success, emotional stability, and strong social skills.

Perhaps you have a child or know children who can't seem to play on their own, free from structured activities. This is the book is for you. Or maybe you have a child who is overactive and having a hard time focusing in school. This book is also for you. If you get constant reports from teachers that your child is not paying attention, or that the school wants to test your child for developmental delays or disorders (such as attention deficit/hyperactivity disorder), this book can help.

Maybe you are getting reports that your sweet and considerate child is starting to push others with too much force during games of tag at school. Perhaps your child is constantly spinning in circles or is so clumsy that she falls or bumps into things, is acting overly silly, or is getting in trouble frequently. Maybe your child worries a lot or is easily upset in new situations.

You are not alone.

An alarming number of children are presenting with these above-mentioned problems. More and more teachers and parents everywhere are reporting that children are starting to fall out of their seats in school, are becoming more aggressive and easily frustrated, are having trouble paying attention, are showing more anxiety, and are spending less time in imaginary play than ever before. These symptoms are due in part to underdeveloped motor and sensory skills, which leave children underprepared for academics and overwhelmed by daily life and social situations. Compromised sensory and motor development can lead to a slew of problems and are quickly becoming an epidemic of grave concern.

Fortunately, there's hope! Scientific and anecdotal research suggests that most of these behaviors are the result of not spending enough time in active free play outdoors. You can solve—and prevent—some of these problems by letting your child play freely and independently in nature. This book will show you how.

HOW PLAYING IN NATURE CONTRIBUTES TO HEALTHY CHILDREN

Active free play outdoors is a kind of play that promotes healthy sensory and motor development in children. It is the antidote to your child spending hours sitting indoors and staring at screens, and to you as a parent being too busy and overscheduled with kid activities to enjoy parenting. The outdoors awakens and rejuvenates the mind and engages all the senses at once.

In nature, children learn to take risks, overcome fears, make new friends, regulate emotions, and create imaginary worlds. It's important that adults allow children both the time and the space to play outdoors on a daily basis. It's important that we give them the trust they deserve and the freedom they need to try out new theories and play schemes.

This book will reveal the therapeutic importance of outdoor play and will equip you with a multitude of ways to foster healthy child development through outdoor play experiences. It is my belief that the developmental benefits of outdoor play can be brought to both the school and home environments. Children can thrive in both settings if they are given the opportunity for free play outdoors *every* day.

WHAT THIS BOOK CAN DO FOR YOU

I wrote this book primarily for you, the parent. If you're like others picking up this guide, you may be feeling worried about, over-whelmed by, confused by, or frustrated by—or maybe all of the above and more—what is happening with your child. But there is hope! In chapter one, I outline all of the developmental changes that are taking place in children, everything from the growing number of children who are being diagnosed with attention deficit/hyperactivity disorder to the increasing number who are falling and getting hurt. Then in chapter two, I jump into the underlying reasons why more and more children are presenting with sensory issues (for example, intolerance to touch and loud noises), attention difficulties, and a variety of other alarming problems. In the remaining chapters I talk about the importance of active free play outdoors; the effect that spending time in nature has on child development, even that of babies; and ways to incorporate outdoor play into the home, day care, and school environments to foster strong, capable, and creative children. By the end of this book you will have numerous strategies to get your child playing independently outdoors whether you live in a rural area, a city, a suburban house, or an apartment.

I also wrote this book for the professionals and educators who work with children on a regular basis. This book sheds light on the majority of behaviors (for example, inattention, fidgeting, clumsiness, lack of creativity, aggression) that many teachers and professionals are observing, explains the origins of these behaviors, and shows how playing outdoors addresses all of these concerns. It also offers solid ideas to get you using the great outdoors as a therapeutic tool to promote independence and creativity in children—preparing them for lifelong learning.

I hope you enjoy reading this book as much as I enjoyed writing it. My wish is that this new knowledge not only inspires you but also spurs you to action. For in order to create change and help the youngest of our society, we all need to do our part.

CHAPTER 1

Why Can't My Child Sit Still?

A perfect stranger pours her heart out to me over the phone. She complains that her six-year-old son is unable to sit still in the classroom. The school wants to test him for attention deficit/hyperactivity disorder. The mother goes on to explain how her son comes home from school every day with a yellow smiley-face sticker for bad behavior. The rest of his class goes home with green smiley faces for good behavior. Every day her child is reminded that his behavior is unacceptable, simply because he can't sit still for long periods of time.

The mother starts crying. "He is starting to say things like 'I hate myself' and 'I'm no good at anything.'"

As I listen to this mother tell her variation of an all-too-familiar story, I can't help but think that this young boy's self-esteem is plummeting all because he needs to move more often.

Perhaps this story hits close to home. Maybe you have a child who continuously fidgets or is disruptive in class. Perhaps teachers encourage you to have your child tested for developmental delay or a disorder. These scenarios can be incredibly tough on parents, leaving them deeply concerned, maybe even worried or anxious about what's going on inside their child. You are not alone. Just thirty years ago, a child being diagnosed or labeled with a developmental disability or neurological problem was very rare. Now it is a growing trend—a trend that should raise major red flags. More and more children are having difficulty with poor attention skills, controlling emotions, balance, decreased strength and endurance, increased aggression, and weakened immune systems. Paralleling this rise in so-called developmental delays is a steady increase in the number of children who need occupational therapy services to treat these issues (Harris n. d.).

This sobering fact is something I've been witness to personally and professionally during the past decade. On the home front, many of my oldest daughter's friends received regular occupational therapy services in their early years, and my youngest daughter needed mild treatment for her own sensory issues. At work, the occupational therapy clinic had a wait list that went out for at least a year.

Within this timeframe, a fifth-grade teacher at my daughter's school told me that her students were having trouble focusing and were fidgeting constantly throughout the day. At her request, I agreed to observe her class. She was reading the children a story from a chapter book as I quietly took a seat and started scanning the room.

It was near the end of the school day. All but one child was fidgeting. Some children were aggressively slapping their wrists, one was continuously rocking in his chair, and one was chewing on the end of a water bottle. Another girl was hugging herself and rocking, while still others tipped their chairs back at extreme angles. Never before had I seen this much excessive movement in a classroom at this grade level. Their fidgety behavior seemed more typical of what one might see in a preschool classroom, not in fifth grade.

Why are these kids struggling? I thought to myself, not just about these fifth graders but seemingly the majority of kids today.

Are we simply more sensitive to children's needs these days? Or is there really an increase in sensory issues in young children? What is causing these problems? I had more questions than answers.

If you've picked up this book, I'm willing to bet that you have more questions than answers too. You've come to the right place for answers. This book is a culmination of more than a decade of observing children, conducting my own studies, and poring over scientific research. The good news is that many of the problems we see in children today are treatable—and in some cases preventable. By the time you finish this book, you'll be well equipped with strategies for ensuring your child is exposed to ideal settings for the kind of active free play outdoors that helps offset developmental problems.

But first, let's take a closer look at prevalent developmental problems facing children today. I'll answer some the most common questions posed by parents and teachers.

DOES MY CHILD NEED THERAPY?

If you are reading this book, there is a strong possibility that your child or a child you know has been referred to occupational therapy, speech therapy, or physical therapy. In fact, according to one study (Szabo 2011), this is true for one in six parents! It's a growing phenomenon that I've been determined to get to the root of. Data from the US Department of Education shows that between 1991 and 2001, the number of five-year-olds receiving what's known as "related services" (this includes occupational therapy, physical therapy, and speech therapy) under the Integrated Disability Education and Awareness Program increased 31 percent. The number of four-year-olds increased 76 percent. The number of three-year-olds increased by 94 percent (Szabo 2011).

According to a study released in the *Journal of Pediatrics* in 2011, one in six children now have a developmental disability. Between 1997 and 2008, there was a 17 percent increase in the number of diagnoses (Szabo 2011). Even at the preschool level, there has been a sharp and steady rise in the number of children who need early intervention—something that was unheard of in years past.

Occupational therapy, among other services, continues to meet the increased demand. A doctor refers a child to occupational therapy services when there is a problem with attention, balance, strength, coordination, or sensory processing. *Sensory processing* encompasses anything having to do with the senses. Common sensory problems in children range from lack of spatial awareness to not listening to not tolerating going barefoot. Occupational therapists help children learn to tolerate a variety of sensory experiences and to maximize their functional independence. Over the last four years, New York City public schools have seen a 30 percent increase in the number of children being referred to occupational therapy. Chicago public schools have seen a 20 percent increase in just three short years. In Los Angeles, the number of referrals has jumped to 30 percent in five years (Harris 2015).

Alarmed by this research, I set out to do my own study. I interviewed ten seasoned elementary-school teachers from different New England states. Because each had been teaching for thirty or more years, I knew that I could get a good perspective on how kids have changed over the past few decades. Across the board, the teachers had similar complaints of their student body. Over the years, they had noticed a slow decline in gross and fine motor ability, safety awareness, self-control, attention, and coordination. Their eye-opening observations are sprinkled throughout this chapter to give life to the data I've described, and a new dimension to the sensory and motor problems you may be seeing in your child and other children.

These teachers' remarks fueled more questions: What is causing so many young children—even toddlers—to require occupational therapy services? Why are developmental disabilities and delays on the rise? Why are more children having trouble with balance, motor skills, attention, and emotional control? What's happening inside our children's bodies? The remainder of this book will explore the answers to these questions—and others posed in this chapter—in detail, offering you peace of mind and an optimistic outlook for your child.

WHY CAN'T MY CHILD PAY ATTENTION?

Over the past decade, there has also been a rise in the number of children being coded as having attention issues and possibly attention deficit/hyperactivity disorder (ADHD). Perhaps your child or someone in his or her class is one of these children. According to a study published in the *Journal of the American Academy of Child and Adolescent Psychiatry* (Visser et al. 2013), an astonishing two million children in the United States were diagnosed with ADHD over an eight-year period (2003 to 2011).

A local elementary-school teacher tells me that at least eight of her twenty-two students have trouble paying attention on a *good* day. "Kids have changed since I started teaching," states Fran Farmer, a highly respected third-grade teacher and one of my interview subjects. I had asked her whether children today appear to be having more trouble focusing and paying attention in class than in years past. "The number of kids with special needs and services has increased. I'd say six out of twenty-two of my kids have a great deal of trouble paying attention. I've had to change the way I teach in order to accommodate this new generation of kids. I used to be able to teach the whole class together as one big group. Now I do a lot of small groups and one on one in order to have success with these kids."

Not only are children having more difficulty paying attention, but they are also displaying this lack of concentration through physical means—in the form of fidgeting. Another teacher laments, "Kids just can't sit still today. It's as if the children don't even care to listen. They are fidgeting all the time and constantly getting up to go to the bathroom." A middle-school teacher states, "They just don't stop moving and getting up from their desks. Every two minutes, they find an excuse to get out of their seats. Sometimes it is to go to the bathroom. Other times it is to sharpen their pencils. I don't remember this happening when I was in school."

In other words, children are doing everything *except* sitting quietly and paying attention. Why are children not paying attention? Why all the movement? Aren't children getting enough time to move around during recess? Doesn't all this movement negatively impact a child's ability to learn and the teacher's ability to teach? What is at the root of all the fidgeting?

WHY CAN'T MY CHILD PHYSICALLY KEEP UP?

Perhaps you've noticed that your child can't quite hold on to and use the monkey bars like you did as a kid. His grip might last a second or two, and then he falls, whining in frustration as he gives up and tries something else. Or maybe you've observed that your child has a hard time walking up a few flights of stairs or a short hill without complaining. This is becoming the norm. Studies and standardized testing are starting to show that the overall strength of children is decreasing. A study published in the child health journal *Acta Paediatrica* looked at how strong a group of 315 Essex ten-year-olds were in 2008 compared to 309 children of the same age back in 1998. The researchers found that the number of sit-ups ten-year-olds could do declined by 27.1 percent. Arm strength fell by 26 percent and grip strength by 7 percent. While one in twenty children in 1998 could not hold their own weight when hanging from wall bars, one in ten could not do so in 2008 (Campbell 2011).

Recently, I observed this decline in strength in a very real way. The woods at TimberNook were alive with play and laughter. It was magical. Kids were collaborating in one corner of the woods with their newly appointed "leader," who distinguished herself with a special feathered mask and long trailing cape. Another cluster of children were building a store using logs, bricks, and rope. A fire was going, and a few kids were roasting biscuits.

All was going well in the TimberNook world, until…

An eight-year-old boy let go of the rope swing in midair and fell to the ground, which knocked the wind right out of him. I calmly but quickly rushed to the boy and quietly knelt next to him. His lips were turning blue. He was panicking. Losing his breath must have been a new and scary sensation.

"Breathe," I gently told the boy. "You are going to be okay. You just lost your breath for a few seconds."

He started crying.

"You are crying. That is a good sign. It means you are breathing."

When he started crying harder, I began to imagine worst-case scenarios as counselors, parents, and children watched with heightened concern. A few long minutes passed. Then, suddenly he got up with relative ease and wiped the tears off his face and the dirt off his pants. He was back on his feet again. I sighed a breath of relief.

This boy had just learned, albeit the hard way, what every child must learn: how to assess one's strength and abilities. It is simply part of childhood. And yet the incident had taken me by surprise; a boy his age and size shouldn't have had difficulty hanging on to the rope. I barely had time to think over the event before three more children fell from the rope swing that week! Falls like this had been really rare. The rope swing is safe, sturdy, and goes all the way to the ground so children of all ages can easily grab hold of it. However, in order to hold on to the rope swing, children must have a strong core, upper body, and grip. The problem isn't the rope swing. The problem is that some children don't have adequate strength.

Children's decreasing core strength should be of particular concern. In 2012, I tested students at a local elementary school on their ability to hold a superman position (on their belly with arms and legs raised), supine flexion (holding an adapted crunch position on their backs), and the plank position when facing the ground. The majority of the children could not meet the baselines taken from the average core strength of children in the early 1980s. The three tested classrooms failed to demonstrate adequate core strength compared to children thirty years ago.

After speaking about TimberNook at a conference for occupational therapists in Ohio, a colleague approached me. She told me that some of the makers of standardized tests to measure strength are contemplating renormalizing the tests because kids are not testing the same as they used to. She went on to say that this was stirring great controversy within the world of occupational therapy. On one hand, occupational therapists want to compare today's children to more up-to-date averages for strength. However, since children are growing weaker, she wondered if we'd be better off holding the kids to the same standards kids were held to in the early 1980s?

This wasn't the first time I had heard this. At a continuing education class for health care professionals, another occupational therapist griped about the same issues. She observed that children were getting significantly weaker. She noted that our society changed pants sizes to accommodate a more obese society, and she was worried that we were going to do the same thing with standardized testing for children—change the "norm" to accommodate weaker children.

Children are performing at a lower level in regards to strength than ever before. In response to this alarming discovery, it may be easier to accept this as a new norm instead of actively working toward a solution. However, when we expect less from our children—instead of holding them to a higher standard—we could be setting them up for failure. Why are our children getting weaker? What does this mean for long-term growth and development? What's at the root of this problem?

Poor Posture Is the New Norm

With decreased muscle strength and lots of sitting, children develop poor posture. I saw this phenomenon firsthand while observing a local middle school. As the teacher lectured to the children for about an hour, their posture grew increasingly worse. Only about one-third of the children had poor posture at the beginning of the period. By the end of the class a good three-quarters had poor posture. Some were hanging over their desks at that point. Others were slouched way back in their chairs. When they stood up, I noticed a few maintained poor posture: a rounded back and anterior head carriage (heads slightly forward).

With a weakened core comes less stability surrounding the spine, which means trouble maintaining good alignment. Dr. Faria, a well-respected chiropractor in my community, reports that 30 percent of her clients are children. She states that many of the children are having trouble "holding" their adjustments and feels strongly this is due to unbalanced musculature. Unbalanced muscles are like pulley systems. If one side is weak, the other side is likely to be tight. For instance, if a child has weak quadriceps (muscles on the front of the leg), the hamstrings (on the back of the leg) are likely to be tight. This can cause pain and poor alignment.

Most of her pediatric clients consistently need to be adjusted at the C1 and C2 joints (located in the cervical or neck region). Tightness in these regions can impinge the nerves and affect input going to and from the nervous system. Having impinged nerves is like taking a garden hose and bending it so that the flow of water slows down a little. When a nerve is impinged or restricted, nerve impulses aren't able to travel to and from the brain as quickly, making response time slower for children.

Have you noticed your child experiencing tightness, especially around the neck region? Tight muscles around the neck and head can be caused by poor posture, by having one's head constantly tilted forward to look at electronic devices, by everyday stress, by restricted movement, and by carrying backpacks that are too heavy.

"Nerves that are impinged or restricted by tightness in the upper neck can affect everything," Dr. Faria states. "Restrictions in the upper neck can affect the eyes, sinuses, and nasal palate—some children may even complain of headaches. Children may have trouble with their pincer gripping from restrictions in their lower neck. Restrictions, regardless of the region, can interrupt the adequate neural input to and from the brain."

A local physical therapist agrees that children's postures are changing. In the past decade she has also seen an increase in back pain with this population. She traditionally works with adults who suffer from chronic back pain, but she is starting to get more referrals than she would like for pediatric clients. She sees children as young as ten years old!

She attributes the sudden increase in pediatric back pain to the hours of mandatory sitting, overall decrease in muscle strength, and the heavy backpacks children are expected to lug around. Many children are presenting with an anterior head carriage, shoulders drawn inward, and abnormal curvature of the spine that puts added stress on the back and neck muscles, causing headaches and pain.

Children can see a chiropractor for back pain, but how do we prevent it in the first place? What are kids not doing that leaves them with poor core strength? Is there a link between poor posture and poor academic performance?

Decreased Stamina

A combination of poor core strength and sedentary behavior can lead to decreased stamina for any type of active play. You many have noticed this with your own child. Maybe she has trouble hiking or needs to take incessant breaks. Or possibly she complains of sore legs after shopping with you for a few hours. In fact, many teachers report that children are increasingly demonstrating poor endurance for physical activity. They state that more and more children complain of being out of breath, that their legs are tired, and that they need to take rest breaks when they go on short nature walks or participate in physical education class.

We often see this at TimberNook, where we have a short walk to get into the outdoor classroom. It is slightly inclined the whole way with mildly rugged terrain, challenging the children to watch where they are going. It only takes about two minutes to walk to the classroom. At the beginning of the week, we often hear lots of moans and groans as the kids make the walk with their stuffed backpacks bouncing up and down on their backs. "This is hard!" a child announces. "When will we ever get there?" complains another. "My legs hurt," a third child whines. It takes a whole week of exposure before the children start tolerating and enjoying the walk.

With any new activity, a different use of muscles coupled with a lack of stamina will make anyone feel some strain. But a two-minute uphill walk shouldn't cause pain or tiredness. Why do children who are hyperactive or fidgety in class complain of leg pain or of being tired when they finally get to be physically active? Is this a catch-22? What's the cause?

Frail Like Your Grandmother's Fine China

Children today remind me of my grandmother's fragile china collection, which was used only on very special occasions, such as Christmas…if we were lucky. Most often, it was the adults who got to use the chinaware, leaving us kids with cheap knockoff plasticware made to look like the real thing. The grown-ups were afraid we'd drop the china and it would break into a thousand different pieces. When I think of the children of today, this image often comes to mind, because if they fall something is bound to break.

Katy Bowman, biomechanical expert at the Restorative Exercise Institute, feels dislocations are now more likely among children due to the combination of heavier body weights and decreased muscle strength. For instance, when a young child tries to hang from the monkey bars and doesn't have enough muscle strength to support his body weight, this load is then transferred to the ligaments, often leading to a dislocation of the elbow. This is what we typically call a nursemaid's elbow (Crawford 2013).

Has your child experienced a fracture or a broken arm or leg? This is becoming more common. The number of fractures in children has increased in recent years. For example, according to a 2010 study in Sweden, the incidence of fractures in children increased by 13 percent between 1998 and 2007. The most common fracture site was the distal forearm, or the region closest to the hand. The most common reason for injury was falling. The study concluded that the increase in incidence was partly due to changes in children's activity patterns (Hedström et al. 2010). Fractures are caused by a combination of factors. Namely, children are more susceptible to fractures when they don't have the muscle strength to protect their bones, as well as when they have weak or porous bones.

When children don't experience enough movement opportunities to challenge and strengthen their bones, the load-bearing capacity of bones decreases significantly. This reduction leads to breakdown and the release of calcium, which is reabsorbed by the body, leaving bones more brittle and weak and increasing the risk of fractures (National Space Biomedical Research Institute n. d.).

Research conducted at the Cincinnati Children's Hospital suggests that millions of children in the United States are not building bones as strong as they should be, which may leave them vulnerable to fractures, rickets, and other bone conditions. Tendons and ligaments can also be affected by lack of activity. When connective tissue remains in a loose state consistent with nonuse, it will gradually shorten and become tight. Tight ligaments, tendons, and muscles are more prone to tears (Southern Illinois University School of Medicine 2007).

Isn't proper nutrition—especially calcium—all that children need to build strong bones? Isn't drinking milk every day enough? How can we expect our children to play if they are at increased risk for fractures? What kind of exercise is appropriate?

WHY DOES MY CHILD FALL SO OFTEN?

While observing a classroom of first-graders, I overheard the teacher saying, in a rather exasperated way, "Okay, we've had a lot of falls this week. Please try to sit in your seats properly." The noise level in the classroom was loud, and the children appeared to be unorganized and excessively silly. Many kids were sprawled across their desks, and others were constantly getting out of their seats. The teacher complained that she was lucky if her kids sat for fifteen minutes.

Many teachers are finding that kids have a lot of trouble with spatial awareness, which leads to clumsiness and falls. A few teachers report that children are consistently draping over their chairs and desks, sometimes even falling out of chairs and onto the floor by accident. In fact, this is happening on a daily basis for many of the teachers I interviewed. Children are even running into the furniture, each other, and sometimes even the walls!

A local middle school had such a severe problem with kids running into each other between classes that the administration placed tape down the middle of the hallways. The kids were instructed to stay to the right of the line while walking, as if they were driving cars.

If you have a child who has difficulty with balance and coordination, you're certainly not alone. In my clinical practice, a major complaint from parents has been that their children are clumsy and constantly tripping over their own feet—whether simply walking across the room or playing sports. For instance, during my daughter Joelle's first few years of hockey practice, it was common for many of the children to be constantly falling and running into the walls and other children, increasing the chance for injury. Now when my daughter plays hockey and effortlessly glides across the ice, not many children can keep up with her, let alone catch her. Her athletic ability, which seems totally appropriate for her age and not advanced at all, is now considered rare.

Research points to a clear and alarming rise in the number of injuries among young athletes in the past fifteen years. For instance, the Center for Injury Research and Policy at Nationwide Children's Hospital completed research on the number of injuries that dancers incur each year. The study found that between 1991 and 2007 there was a 37 percent increase in the number of dance-related injuries for children (Nationwide Children's n. d.). The most common cause of these injuries was falling.

Even more alarming is the number of injuries occurring during physical education (PE) class. One study found a 150 percent increase in the number of PE-related injuries for elementary, middle, and high school students in the United States between 1997 and 2007. "It is unlikely that this increase was attributable to an increase in PE participation," explains study author Lara McKenzie, PhD, principal investigator at the Center for Injury Research and Policy and faculty member at the Ohio State University College of Medicine (Nationwide Children's 2009).

What is contributing to this significant rise in the number of kids falling? What is making today's kids clumsier than kids of a generation ago? If children can't walk down a hallway without bumping into others or play sports without getting injured, what does this say about their overall physical and neurological health?

WHY DOES MY CHILD HAVE AN ENDLESS COLD?

It is now a prerequisite for every child at my daughter's school to bring four boxes of tissues at the beginning of the school year. *How could she possibly need four boxes of tissues?!* I thought to myself the first time I was asked to donate. But sure enough, by the end of the year all of the boxes had been used. Some teachers even asked parents to donate more boxes.

Perhaps your child is constantly sick during the winter, getting one cold after another, missing school and getting other siblings—or even you—sick in the process. This is an unfortunate reality shared by countless parents I've talked to. They say things like, "It seems like he has one big cold that lasts all winter." During my daughter's hockey games, some children need tissues constantly while playing. I asked one mother if her child had a cold. "No," she said. "That is just Sarah. Her nose constantly runs during the hockey season."

Besides the obvious health issues that come with sedentary behavior, such as obesity and an increased risk of diabetes and high blood pressure, children also develop weakened immune systems. They are more prone to colds, illnesses, and allergies. The global incidence of childhood allergies, asthma, and eczema increased by 0.5 percent annually between the mid-1990s and 2002, according to the International Study of Asthma and Allergies in Children (Asher et al. 2006).

David Brownstein, a board-certified family physician and one of the foremost practitioners of holistic medicine, talks a lot about the immune system. He tells a story in one of his articles about being approached by a fan in an airport. She asked him, "Why are so many kids having all these allergies? We never saw peanut, milk and gluten allergies when we were kids—where is it all coming from?" He answered, "I believe that there are multiple reasons for this, but the main reason is that the younger generation's immune system is becoming weaker and weaker" (quoted in Hubbard 2005).

Why do children's colds seem to last forever? What is causing the increase in allergies and asthma? Is this increase mainly related to food and the environment? Or is there another factor to consider?

WHY IS MY CHILD SO AGGRESSIVE?

"Tag"—a cherished game that you most likely remember playing as a young child. It seems innocent enough. But is it? Not according to many teachers.

What was once considered a simple and honest game of good fun has become a nightmare on the playground. Children are starting to hit with such force that they often end up whacking their opponent across the back with a monstrous slap. I've seen this myself at TimberNook.

"Ouch!" one kid cries, now on his hands and knees and fighting off tears. "Don't hit so hard!"

The child standing over him counters, "I didn't mean to…"

The act seems unintentional, though painful nonetheless. Aggressiveness during tag has become such an issue that schools across the United States are starting to ban this once beloved game.

In the fall of 2013, the problem with this classic game hit close to home. At a local New Hampshire school, tag was banned due to safety concerns. Parents and children were confused, and some were outraged. Headlines stated everything from "Banning Tag at Recess is Dumb" (Stevenson 2006) to "More Schools Banning 'Tag' Because of Injuries" (Wang 2013). Curious, I asked local teachers what they were observing at recess time.

As I suspected, teachers were seeing increased aggressive behavior during recess. One teacher stated, "We have to make up extra rules for them because they have trouble knowing how to use appropriate touch with one another." (Before the ban, administrators had implemented a two-finger touch rule to prevent kids from pulling other kids down.) Another teacher in a different state griped, "They can't keep their hands off each other! We speak to them but they can't seem to help it."

It appears that children are becoming more aggressive. Do children today simply not understand how to play safely? Are they growing up exposed to things that make them mean-spirited? Or is it that kids don't have good body awareness?

WHY DOES MY CHILD HAVE DIFFICULTY READING?

Myopia, or nearsightedness, is on the rise. It is more common in children today in the United States and in other countries than it was in the 1970s. In fact, myopia is nearly reaching epidemic proportions in parts of Asia. For example, in Taiwan the percentage of seven-year-olds suffering from nearsightedness increased from 5.8 percent in 1983 to 21 percent in 2000. Furthermore, an astonishing 81 percent of Taiwanese fifteen-year-olds are myopic (Palmer 2013).

Curious, Kathryn Rose, an orthoptics professor, decided to take a closer look. She found that only 3.3 percent of six- and seven-year-olds of Chinese descent living in Sydney, Australia, had myopia compared with the 29.1 percent living in Singapore. Even more interesting is that time spent in front of a screen or reading a book couldn't account for the discrepancy (Palmer 2013).

Though myopia isn't as prevalent in the United States as it is in Asia, the number of cases is rising quickly. A 2009 study showed that myopia among Americans between ages 12 and 54 swelled from 25 percent in the early 1970s to 42 percent around the turn of the century (Palmer 2013).

Not only are many children having trouble seeing things from a distance, but with the rise in developmental disabilities we are seeing today, it makes sense that we'll see a rise in visual deficits as well. Is your child having vision difficulties? Has your child complained about headaches or an inability to read or write at the same level as peers? One problem that many therapists are seeing today, as opposed to thirty years ago, is that more and more children have trouble using the muscles of their eyes in unison, say to scan a room to find an object or to read a book accurately.

Oftentimes, these vision problems go undetected, and children struggle in all aspects of their schoolwork. Typically, schools only assess a child's ability to read letters or numbers off a chart. This tests their visual acuity. However, it is rare for schools to assess children's ability to track and scan and effectively use their eye muscles.

A skilled reading specialist who works for a local charter school and one on one with many children agrees that there is a growing problem with visual difficulties. She believes that many of the kids she sees for reading intervention have some sort of difficulty with their visual skills.

I once treated a little girl for sensory and motor development issues. She constantly held her hand over one eye in school in order to try and make sense of the words in front of her. She was in the first grade and could not yet read. The elementary school that she attended tested her ability to read an eye chart, and she passed with flying colors. "Her vision is fine," the school reported. However, no one thought to look more closely at how her eyes were working. I assessed her and quickly realized that she couldn't get her eyes to move from point A to point B without them looping in circles. No wonder she had trouble reading!

Why is myopia on the rise? Why are children having trouble coordinating their eye muscles to scan a room or read a book?

WHY IS MY CHILD SO EMOTIONAL?

"I love camping!" said six-year-old Joelle to another girl her age as we were walking in a big group. "We go in a tent. Do you go camping?"

The little girl looked at Joelle with disgust. "Camping? With all those bugs? Ticks could be crawling under your tent. No thank you!"

Joelle then noticed a perfectly manicured patch of grass and laid down on it. "Look at this grass!" she squealed with delight.

"Get up!" the other girl panicked. "There may be ticks in that grass! Ugly, awful ticks. Hurry, get up!" The girl was stricken with fear.

It appears that more and more children are having trouble controlling their emotions, while anxiety issues are at an all-time high. Worries and fears are taking over and becoming a barrier to enjoying the simplicity of childhood. Why are worries and fears increasing in children? Why are children having trouble controlling their emotions? Are we just becoming more sensitive to these issues? Or is there a neurological reason for the upsurge?

Trouble with Emotional Control

Teaching children how to *self-regulate*, or control their emotions, is a hot topic these days. In the past five to ten years, I have noticed a substantial increase in the number of blog posts, articles, and self-help books focused on teaching parents how to help their children learn to effectively self-regulate. Clearly this is becoming a big enough issue that we now need an array of strategies from which to choose. Tactics such as yoga, meditation, and modeling are often employed. Something that should be intuitive and a natural form of development now needs to be taught.

Joleen Fernald, PhD, a speech language pathologist whose specialty is working with children with anxiety-based disorders, has noticed a rise in social-emotional issues in the past ten years. "Even eight-year-olds can't self-regulate!" she says. "They should be able to start doing this at three months of age. I'm not referring to children on the autism spectrum. These are the kids who don't have a diagnosis or receive special services."

The teachers I spoke with say that children today appear to have more trouble controlling their emotions in the school environment than children did thirty years ago. They comment on the fact that many children are quick to cry in class or get easily frustrated. One teacher states, "In the past, it was rare for a child to break down and cry in the middle of class. Now it is a frequent occurrence. Even more surprising is that many of them are boys." They also notice that kids seem to have less interest and motivation when it comes to school projects and lessons. Many parents (perhaps you are one of them) are becoming increasingly concerned because their children are coming home in the early grades saying they hate school. Not the best way to start one's academic career.

What is behind the increase in mood swings? Why can't children today seem to self-regulate like generations before?

Rise in Anxiety

Does your child suffer from anxiety? Maybe your child is scared of the dark or afraid to get sick. Maybe he or she has difficulty in new environments or worries about getting hurt. According to one study (Cohen 2013), at least one in four of you reading this book will have a child with a diagnosed anxiety disorder! Licensed psychologist and award-winning author Lawrence Cohen, PhD, states, "I believe that childhood anxiety is indeed on the rise at every level, from fears of monsters under the bed to phobias and panic attacks to severe anxiety disorders." He goes on to explain that when he was training to be a psychologist thirty years ago, 10 to 20 percent of children were born with a temperament that was highly reactive to anything new and unfamiliar. Some of these children went on to be anxious, timid, or shy later in life. A much smaller number, about 1 to 5 percent, were actually diagnosed with a full-fledged anxiety disorder at that time (Cohen 2013).

Nowadays, 10 to 20 percent of children still have that reactive temperament, but the number of children diagnosed with an anxiety disorder has skyrocketed to 25 percent according to the National Institute of Mental Health (Cohen 2013). We see symptoms of anxiety on a regular basis at our TimberNook nature camps. Our first hint that a child may be anxious comes from parents who ask about what strategies we use to transition children into the camp environment. "He doesn't do well with change," one mother states. Another says, "She has a hard time transitioning to new situations."

We also get many children who have a clear fear of something outdoors. For instance, they may be afraid to go in the forest due to lack of exposure. Or they make it very clear that they don't want to take their shoes off at the beginning of their first week at camp. Others are afraid to touch the chickens for fear they may get pecked. Approximately five out of twenty children who come to camp each week have some form of anxiety.

Are the brains of our children wired differently than those of children twenty years ago? Or are there environmental factors at play? What's causing so much anxiety? Moreover, how can we prevent some of the symptoms of anxiety from occurring in the first place?

WHY DOESN'T MY CHILD LIKE TO PLAY?

Have you sent your children out to play only to have them return minutes later saying, "There is nothing to do out there," or "I'm bored"? Studies indicate that children's play habits have changed drastically in the past few decades (Bundy 1997; Juster, Ono, and Stafford 2004). The amount of time children spend in unstructured play has decreased by 50 percent, resulting in children devoting most of their time to indoor activities (Clements 2004). Children are also spending more time than ever before in front of electronic screens. Research suggests that the average child spends at least six hours a day in front of the television or a computer or playing video games (Rideout, Foehr, and Roberts 2010). With the substantial decrease in the amount of time children spend in unstructured play outdoors, it is no wonder that they have trouble playing independently and creatively.

Teachers often take turns observing children at recess time. When I asked a few teachers about the play skills of children today compared to children thirty years ago, here is what one said: "There is less imaginary play. We used to see a lot of 'pretend play' in the past—children creating their own games and worlds on the playground. Now, they gravitate toward the play structure or play a game of tag until the whistle blows to go back inside." She went on to say, "It is *really* noisy and crazy. It seems like they run around without a purpose. There is little creativity like we observed in the past. It is like they don't know what to do with themselves. There are a lot of children tattling and coming up to us to seek constant direction and reassurance on what to play or do. It is both frustrating and sad to watch."

Children are starting to lose both their desire and their ability to play—something that should be fundamental to human nature. Why aren't many children developing their own games, if they want to play at all? Why do they seem to prefer structured activities to unstructured ones and to seek adult guidance rather than lead peers on their own? Is the lack of imaginative play harming them in other ways?

IN A NUTSHELL

The cold hard truth is that when you compare today's children to past generations, they just can't keep up. Children are getting weaker, less resilient, and less imaginative. They're having trouble paying attention in school, experiencing difficulty controlling emotions, and having trouble safely navigating their environment.

Nutrition and exercise can help the above-mentioned problems, and these are two common areas that most parents focus on. It has been proven that addressing nutrition and exercise fights obesity and improves overall health and academic performance. However, focusing solely on nutrition and structured exercise as the solution appears to have contributed to the decline in sensory and motor development we see in children today by overshadowing our understanding of these important developmental milestones.

The good news? There is hope for your child! The answer is really quite straightforward: active free play—ideally outdoors—is essential for your child's sensory and motor development. Allowing your children time and space to play outdoors on a daily basis can significantly improve and encourage healthy development. The following pages will spell out how to do this. So what about all the questions we raised in this chapter? Read on to find the answers.

CHAPTER 2

The Body and the Senses

Before we explore answers to the dozens of questions raised in chapter one, let's first take a close look at our children's bodies. In fact, let's take this opportunity to marvel at the tremendous amount of growth their bodies experience. Their bones and muscles never stop growing, their senses never stop sharpening, and their neurons never stop firing. It's quite remarkable how your child can go from having no muscle control to walking and possibly running in just one year. And in another year, your child can go from standing on one foot to riding a tricycle independently.

Children literally thrive by challenging their bodies. When their bodies aren't challenged, they fall behind in their development and run the risk of exhibiting some of the problems discussed in the previous chapter. In order to reverse the growing trend of children who have trouble with strength, balance, and coordination, we need to first understand the underlying motor and sensory skills that support and develop healthy, strong, and capable children. So let's examine not only how the body grows but also how your child's environment can impact that growth. We'll also take a brief look at how social and emotional skills develop.

THE BODY

I once met a parent who told me that her two daughters had a daily ritual of climbing an old, fat oak tree just outside their home. According to the mother, the tree was perfect for climbing. The girls would get home from school, throw their backpacks to the side, and up the tree they'd go. When they arrived at TimberNook, they scaled about ten different kinds of trees in our woods— skinny trees, very old trees, trees with branches close together, and trees with branches far apart. Each tree was a celebrated new challenge for them. They swung from branch to branch in an effortless motion, checking the branches as they went to make sure they were sturdy. They climbed backwards and sideways, scaling around the tree as they went. It was through daily tree-climbing practice that these girls developed the necessary skills they needed to go up and down the oak tree back home with such precision and ease.

There are many ways to approach a discussion of the body, but for our purposes we'll look at the development of two primary skills: gross motor skills and fine motor skills. These skills are honed when the muscles, brain, and nervous system work together to allow your child to perform any physical action, from kicking a ball (gross) to drawing with crayons (fine). It takes daily practice to refine and strengthen both fine and gross motor skills.

Gross Motor Skills

Gross motor skills encompass whole-body movements and coordination of the legs, arms, and other body parts in order to walk, run, and climb, among other things. Some of a baby's first gross motor milestones are the most memorable, such as crawling and taking first steps. By the time children reach two years of age, they are able to stand up, walk and run, climb on and off objects, walk up and down stairs, ride on push toys, and even stand on one foot. These gross motor actions are built upon and enhanced through sensory experiences, continued practice, and refinement all the way up to adulthood.

In order for gross motor skills to become precise and accurate, throughout the day children need to practice activating large muscle groups (muscles of the legs, arms, stomach, and back) through a variety of movements and sensory experiences. Let's use walking as an example. When children first learn how to walk, they are very unstable. They may take a step or two and then fall to the floor. They walk with their arms up in the air in order to provide extra balance and support. With lots of practice and opportunities to walk, children start to take more steps. Soon they are taking four or five steps before falling. Next thing you know, children are walking from one side of the room to the other with a faster pace and more solid footing. As they practice, children lower their arms as they become more efficient and stronger and the motor skills become unconscious acts. Less concentration and effort are needed.

But once a toddler becomes a walker, that doesn't mean the skill is mastered. Far more honing and strengthening are still needed. A good way to get young children to continue to work these important muscle groups is to introduce them to the outdoors. The terrain is variable, and there are obstacles, interesting objects to climb, rocks to jump off, and bugs to chase. It is a much more varied environment than a child's playroom or a gymnasium. Offering play and exploration outdoors will challenge your children's motor skills all the way to adulthood.

In addition to leg and arm muscles, core muscles (the stomach and back muscles) and neck muscles are also used for gross motor actions. Although it might not always be obvious that core muscles are engaged when climbing a tree or hopping from rock to rock, they are working hard to keep children upright and balanced.

Being literally in the center of your child's body, these muscles are also central to just about every activity. They, along with all the large muscle groups, provide the foundation and support for the smaller muscles of the arms, hands, and fingers to work effectively. Without adequate gross motor strength, coordination, and control, it becomes very difficult to master fine motor skills, such as buttoning a shirt, cutting with scissors, and taking off shoes.

Therefore, it is important that children are given ample time to develop gross motor strength and skills in order to support all of the other motor skills—both large and small. Read on to learn more about why strengthening these muscles is so important.

THE IMPORTANCE OF BUILDING STRENGTH

Children develop strength when they have daily opportunities to activate and use big muscle groups in a variety of ways. For instance, when babies have plenty of time to be on the ground day after day, they build strength simply by interacting with the environment around them. They reach for objects, attempt to kick things, push up for a better view, and roll over for a new perspective. They don't need to do formal baby exercises that so many parenting forums recommend; simply moving about in a sensory-rich, yet soothing, environment is more than adequate for developing muscles naturally.

Children have an innate curiosity and desire to move. They may crawl over to investigate a new sound they hear on the other side of the house or across the lawn. If something catches their eye, such as a colorful bug crawling on a tree just out of their reach, they may pull themselves to standing to get a closer look. This motivation to learn about the world around them propels them forward, activating the muscles as they go.

The outdoors provides children with a changing environment. The sun coming out from behind a cloud might prompt a crawler to scoot from the shade into the sunlight or an older child to happily spin in circles. A grassy slope might encourage some to roll down it, others to make a race for the top. Engaging the large muscle groups in varied ways on a regular basis builds a strong base of support that allows the arms, legs, head, and eyes to move around and function better, leading to better looking and listening skills and more accurate and efficient body movements.

Without adequate strength, which comes from practice, practice, practice, all things require more effort and conscious thought in order to effectively move about the environment. Weak gross motor skills in children can lead to difficulty sitting upright in school, poor endurance in physical education class, inefficient body coordination, and even injury, as we touched upon in chapter one.

To keep gross motor skills in optimal condition, it's important to ensure that children under two years of age get to move *throughout* the day—preferably a total of four hours or more of active movement. Give them regular opportunities to crawl, climb, jump, roll, walk, and run. Older children should ideally be exposed to three hours a day or more of the aforementioned plus squatting, hanging, carrying heavy objects, jumping off of things, tumbling, and other rigorous activities.

CORE STRENGTH

Core strength is often misunderstood. When most people think of core strength, they think of rock-solid six-pack abdominals, something you might show off in a bathing suit. However, core strength comprises many muscle groups, not just the ones you see on the outside. The core consists of the outer core, or the abdominals, and the inner core, the muscles that surround the hips, pelvic floor, diaphragm, and spine. Having a strong inner core provides stability for the spine and allows for good alignment and fluid movement. The inner core supports postural control and deep breathing patterns and provides a strong base from which the outer core muscles can move. The inner core keeps children upright and stable. When children don't have good inner core strength, they have to compensate by using the more superficial outer core muscles to keep their bodies upright.

Outer core muscles were designed for more refined and quick movements, such as pushing and pulling actions (for example, pushing open a door or pulling yourself up onto a tree limb), not prolonged periods of use! It takes much more energy and concentration to keep the outer core muscles engaged for stability purposes; therefore, if children have a weak inner core, they will tire easily and lose attention more quickly since these muscles require much more conscious effort to engage. Children with poor inner core strength often present with a slumped posture, decreased endurance, and poor balance.

We start to develop core strength as infants. If children are given frequent opportunities to be on the floor as babies, especially on their bellies, they will start to develop their core muscles. For instance, when babies are given "tummy time," they learn to start lifting their head up. This develops the muscles in the neck and back. The neck muscles need to be strengthened in order to support better looking and listening. Tummy time also helps them start to transfer their weight to other parts of their body, strengthening those areas as well. As they are allowed more floor time, they continue to develop the muscles in the back and the stomach—rolling from their back to their stomach and vice versa.

Older children develop core strength through a variety of free play activities—especially outdoors. They challenge and strengthen their muscles as they climb up trees, roll up and down hills, swim, ride a bicycle, swing, jump, and run around. For instance, riding a bicycle requires that you balance and stabilize from the center, or core, in order to coordinate arm and leg movements effectively.

Climbing a tree requires activation of the inner and outer core in order to provide stability and mobility as a child adjusts her body and pulls up and up again. Once more, there is no need for formal exercises. When your children are given both the time and space to move, they'll naturally be getting all the "weight lifting" exercises they need to develop a strong and stable core.

UPPER BODY STRENGTH

When therapists refer to treating the "upper body" muscles, we are generally speaking about the muscles of the arms, chest, upper back, and shoulders. These muscles provide the foundation for the more refined movements of the fingers and hands. For instance, in order to hold and write well with a pencil, a child first needs adequate strength and stability in the shoulders and arms.

Strong, developed upper body muscles also allow for precise, fluid movements of the upper body and arms. They allow us to complete tasks such as swinging a baseball bat with speed and accuracy or crossing the monkey bars with little effort. For babies, crawling is a significant milestone that strengthens the upper body and develops strong shoulders, laying the foundation for fine motor skills, such as picking up small objects and, later, scribbling with a crayon.

Older children continue to refine and strengthen their shoulders through a variety of play experiences, such as practicing cartwheels, holding onto a rope swing, climbing a jungle gym, and playing pass with a large playground ball, to name a few. The more play opportunities children have, the more they strengthen their muscles.

ENDURANCE

"Endurance" means being able to persist through physical challenges. Children with poor endurance tire easily and give up more quickly. Having good endurance is important. It allows children to play for hours without needing to take lengthy breaks. It fosters a healthy body and immune system. It promotes excellent athletic performance. Overall, children feel better and gain more confidence if they can play without needing time-outs.

To build good endurance, two things need to be in place: adequate strength and heart-pumping active play. Children need many opportunities to improve and maintain strength. As we just learned, this can be achieved through plenty of outdoor play. Lifting heavy rocks to build a dam, climbing trees, scaling a rope ladder, digging at the beach, pumping on a swing, and biking are all great examples of children playing and challenging their strength at the same time.

Active play that gets the heart pumping is important not only for health reasons but also to strengthen the heart and the lungs' ability to take in oxygen. With regular active play, such as running through a meadow, playing flashlight tag, swimming in a pool, and skateboarding down the street, children improve the ability of the heart and lungs to support increased activity. Their stamina increases, and they can tolerate even more rigorous play.

POSTURAL CONTROL

Postural control is the ability to maintain body alignment. Adequate postural control is important in order for babies to reach their developmental milestones, from sitting upright to crawling and finally to standing. It helps them maintain their balance and stability as they move around their environment. They achieve postural control as they learn to push against the forces of gravity (Case-Smith 2001).

Over time, and with lots of practice moving around on the floor, babies learn to hold their heads and bodies upright so that they no longer need support when they are carried. They'll easily transition from being on all fours to sitting upright and then to standing if given the chance to move on their own. The wobbliness so characteristic of really young infants starts to go away as they become increasingly sturdy and in control of their bodies. Children continue to develop, strengthen, and finally maintain postural control through continuous play experiences that work against the forces of gravity.

When children move their bodies against gravity they develop the necessary strength and balance patterns needed to maintain postural control. Swinging, climbing up a pole or tree, riding the merry-go-round, winter sledding on the belly with arms and legs up, jumping rope, and rolling down a hill and back up again are all play experiences that help children develop postural control. Contrary to popular belief, continuously sitting upright *doesn't* build good posture—it actually fatigues the muscles, which then leads to poor posture. Having children sit with perfectly straight spines might be constructive during piano lessons, but it's really not wise for prolonged periods.

When children are restricted in seated positions for many hours, such as by baby devices (for example, bouncy seats and infant seats) and by unrealistic rules for older children to sit for long periods (for example, while doing homework and being lectured at during school hours), it is hard for them to develop and maintain adequate strength and control. Children who lack postural control may be more apt to fall out of their seats in school, may need to lean on the table for support while seated, may have trouble getting on and off playground equipment without falling, and may sit or stand with a slouched posture. Therefore, it is important for your children to experience frequent play opportunities that challenge them to move their bodies against the forces of gravity.

GROSS MOTOR COORDINATION

Gross motor coordination is the ability to repeatedly execute a sequence of movements with accuracy and precision. Without good gross motor coordination skills, children are clumsy and more prone to accidents. In order to develop good gross motor coordination, children really need to have good awareness of where their body is in space (called vestibular sense, which we'll talk about in the second half of this chapter) and a strong core. Many occupational therapists talk about the importance of "crossing the midline" (your center) as a prerequisite for developing basic gross motor coordination. Well, the inner core (muscles of the hips, spine, pelvic floor, and diaphragm) establishes the midline! Without good core strength, there is no perception of center—no anchor on which to support smooth and efficient body movements. Therefore, in order to develop good coordination, children need plenty of opportunities to move and strengthen their bodies first.

Once the awareness and strength of the core is well established, children start to experiment with playing at their center. For instance, children as young as six months of age start banging objects together near the center of their body. Later, they start clapping and finally reaching across their body to either crawl or grab an object. These are the first steps toward developing higher-level coordination skills. Children later develop more complex gross motor coordination skills, such as crawling. With crawling, babies learn how to use the arms and legs in alternating patterns—a more complex skill. Each new skill learned lays the foundation for a higher-level skill. Before you know it, kids are kicking balls and jumping rope.

Older children continue to refine their gross motor coordination skills through practice. It's that simple. If children climb a tree once or twice a year, they are likely to remain novices and stay fairly close to the ground. However, if children climb trees on a regular basis, like the girls mentioned at the beginning of this chapter, they'll not only develop the muscles and advance the coordination needed to be expert climbers, but they'll likely be confident, strong, and safe in any other physical pursuit!

Fine Motor Skills

Fine motor skills are those that involve small muscle movements, usually involving the hands and fingers in coordination with the eyes. Fine motor skills are involved in grasping a piece of cereal to place it into the mouth, holding a pencil to write on paper, lacing a pair of shoelaces, and holding a knife to cut meat with precision. In order to have good fine motor skills, children need a strong core and good shoulder stability. They need to rely on this solid base in order to devote concentration and energy toward more precise work. Once a solid base is in place, just like with gross motor coordination skills, children need plenty of practice grasping objects and manipulating them in order to develop strong and capable fingers.

FINE MOTOR STRENGTH

Fine motor strength pertains to strengthening the small muscles of the hands, fingers, and wrists. Fine motor strength is needed to complete many of our daily tasks, such as zipping up pants, turning a key to open a lock, holding on to the monkey bars, opening up packages, cutting with scissors, and so on. Children who have trouble with fine motor strength will have trouble doing all the above-mentioned skills.

Children develop fine motor strength by interacting with small objects or performing tasks that provide resistance. For babies, crawling strengthens and develops the arches in the hands, later needed to grasp small objects. As with the gross motor skills, children need ample time to use the hands to explore their surroundings. They need opportunities to pick things up, turn them around in their hands, open them up, release them, and then move on to the next thing. The more chances your children have to use their hands, the more refined and strong the muscles in the fingers and hands will get. Things like manipulating clay, digging in the dirt with a spoon, pulling weeds in the garden, and writing with chalk on the driveway are all excellent play experiences for kids to naturally strengthen and maintain strength in their hands.

Older children continue to refine and maintain fine motor strength through a variety of everyday activities. Outdoor chores such as shoveling, raking, and digging in the garden require a nearly constant firm grasp as the chore is completed.

Working with tools also requires a lot of fine motor strength. Things like hammering together a fort allows a child to hold and maintain a strong grasp on the hammer while the other hand holds the nail still with a more refined grasp. Using a screwdriver, manipulating nuts and bolts, and even handling a pocketknife to whittle wood all challenge the fingers of the hands. Yes, older children *can* use a knife, and it's an important skill to have. Most injuries happen when hand muscles are weak and fine motor skills are not challenged. A child who has strong hands and arms shouldn't be any more prone to injury than an adult.

FINE MOTOR COORDINATION

When a certain amount of body stability and hand strength has developed, the hands and fingers are free to begin to work on movements of dexterity and isolation as well as different kinds of grasps. As children manipulate different objects in their environment, they develop ever more precise movement patterns. For instance, a baby starts by swatting at an object with his hand (palm open) over and over again. Later he starts grabbing on to toys with a full fist latched on.

After much practice, the hands get stronger and babies become more aware of and able to control the individual fingers in isolation. That is when they start grasping using just a few fingers. Eventually they learn more complex skills, such as holding a spoon to bring food to their mouth without spilling and scribbling with a crayon while the other hand prevents the paper from moving.

A child with poor coordination will have trouble using the hands to do just about anything, from cutting paper to tying shoelaces to coloring. Children develop good coordination skills by using the fingers and hands in new and challenging ways daily. It takes many opportunities of grasping things and manipulating objects in order to fine-tune skills. If your children get tons of chances to develop a strong and precise grasp and fine motor dexterity and control, things like holding a pencil correctly or even playing the piano should come naturally and more readily.

Older children refine and practice fine motor coordination when they have opportunities to use their hands in a variety of ways. Things like braiding friendship bracelets, sketching in a notepad, knitting wool mittens, using a screwdriver to screw and unscrew, creating fairy houses and other miniature landscapes, and working with pottery are all great ways to challenge the fingers and improve coordination. Just like with gross motor coordination, the more practice your children get, the more adept they will be at manipulating and coordinating the fingers.

THE SENSES

We all know the five senses: touch, sight, listening, smell, and taste. But did you know that there are actually two more senses?

Proprioception is the ability to sense what different parts of your body are doing without even looking at them. The *vestibular sense* is your awareness of where your body is in space; it determines your ability to effectively navigate your environment with ease and control. All of our senses affect our ability to function in our environment. In fact, our evolutionary survival depended on them! They can alert the body to danger, help us to remain calm, and even ground us.

Sensory feedback gives us important information about our surroundings. The sound of laughing children, the chill of an approaching storm, the smell of smoke—these things give us information and make us feel and act in certain ways. The act of processing and making sense of the input coming in—bringing together all the puzzle pieces to form a nice picture of our environment, our body, and our body's abilities—is called *sensory organization*. The calmer and more alert we are, the better able we are to process and organize our senses.

On the other hand, *sensory disorganization* occurs when there are too many senses being activated at once. Our body can't decode this information properly, and we often end up having a *fight-or-flight response*. Sensory overload can cause our nervous system to react as if we're in danger, and it prompts us to fight (stay and fight) or take flight (run away).

When we have a fight-or-flight response, we start to experience bodily symptoms such as increased heart rate, dilated pupils, rapid breathing, muscle tension, and increased perspiration. These reactions are normal if we are in real danger, say coming face to face with a bear. However, this response is not helpful if your child is about to take an exam and there are too many bright colors or loud noises in the classroom causing unnecessary anxiety.

Let's look at all seven senses and learn how they work and develop. When you understand their important role in your child's overall development, you can begin to optimize sensory experiences so that he or she receives states of sensory organization and avoids situations of disorganization.

Touch

The tactile, or touch, sense is the very first to develop in the womb and is the largest sensory organ of the body (Biel and Peske 2009). We receive our touch information through sensory receptors, or cells, that are all over our skin, from head to toe. Touch sensations of pressure, vibration, movement, temperature, and pain activate these tactile receptors and bring input to the brain for interpretation. We are always actively or passively touching something, whether that is a light breeze brushing up against us or the cool, hard floor under our bare feet (Kranowitz 1998).

The touch sense gives us necessary information about the environment. It tells us what the temperature "feels" like outside, gives us feedback about the object we are touching (for example, rough versus smooth, cool versus warm, hard versus soft), and indicates if something is painful (such as a scrape on the knee). The touch sense can also alert us to danger. For instance, if your back has been hurting you, you are likely to seek medical attention before further damage is done. It can also comfort us. Getting a warm embrace from mom is often soothing and calming for young children.

Newborns have a rudimentary sense of touch. They can sense when they have a wet diaper and will turn their head reflexively if someone strokes their cheek. However, they cannot yet determine very well where they are being touched. The brain has not learned to differentiate one spot from the other. As children are exposed to a variety of tactile experiences they learn how to discriminate where they are touched and what they are touching, and they learn about themselves and the objects around them.

When children have difficulty with the sense of touch, they may overreact to tactile experiences. We call this *tactile defensiveness*. Children who have tactile defensiveness may experience a fight-or-flight response to objects that aren't necessarily harmful. For instance, they may become easily upset when their hands or face are dirty; they may avoid being touched, dislike brushing their teeth, or become anxious about walking barefoot on sand or grass. Some children are the opposite and are undersensitive to touch. They may not even realize when they have a new scrape, or food on their mouth.

Proprioception

Proprioception comprises sensory receptors in the joints, muscles, ligaments, and connective tissues that tell you where your body parts are without you having to look at them. The receptors sense when muscles and other connective tissues are stretched or at rest (Biel and Peske 2009). Our brain analyzes the information from the receptors and gives us a sense of body position and motion. Proprioception regulates how much force you need to use when completing tasks, such as peeling a boiled egg without crushing it, holding a baby chick without squeezing too hard, and writing with a pen without ripping the paper.

Children develop proprioception through a series of pushes and pulls that happens when they interact with their environment, such as by picking up heavy sticks and putting them back down again to build a fort, raking leaves, and shoveling snow. This push and pull creates new gravitational loads and adaptations that strengthen the bones and muscular tissue over time, offering increased awareness of the different muscles' capabilities and positioning for better body awareness.

Children with poor proprioceptive sense are generally more susceptible to fractures, falls, dislocations, and injuries. They tend to be clumsy and have been known to walk in a robot-like fashion. They often have to look at their body parts in order to move them correctly. They may have a hard time regulating how much force to use when walking, hugging, jumping, and so on. Without proper proprioceptive feedback, children may fall out of seats, fall frequently, and trip while walking up stairs. These children tend to be more accident-prone. Remember the discussion of the game of tag in chapter one? Poorly developed proprioceptive sense is one reason why children are hitting with too much force during tag.

To maintain or strengthen the proprioceptive system, encourage your child to have play experiences that offer resistance to the joints, muscles, and connective tissues. This can also be referred to as doing "heavy work," which basically consists of activities that require pushing, pulling, and carrying heavy objects. Pulling a wagon with another child in it, picking up heavy rocks to build a dam in a stream, and digging in the dirt or the sand are great ways to get nice sensory input to the joints, muscles, and connective tissues. Heavy work helps to develop a strong and capable proprioceptive system.

Vestibular Sense

Remember the fifth-grade classroom I observed in chapter one? The class that couldn't stop fidgeting? Well, a colleague and I decided to investigate a little further into the reasons why the children were constantly fidgeting. We took three classrooms from that art-integrated charter school and tested their core strength and balance skills and compared the results to the average core strength and balance skills of children in 1984. We found that only *one* out of twelve children in the groups we tested had the average core strength and balance that kids had thirty years ago! The data from this pilot study was eye-opening. How could eleven out of every twelve children have such pronounced strength and balance deficits when compared to children in the early 1980s—my generation?

When children participated in simple balance tasks, most had a difficult time. For instance, we asked children to spin in a circle ten times with eyes open and then with eyes closed. In both situations kids fell to the ground; some shuffled their feet at a snail's pace, some had eye responses that were not appropriate (for example, eyes moving back and forth at a rapid pace for longer than typical after spinning), and others extended an arm as a visual guide instead of relying on their body sense. To observe so many children struggle with the simple act of spinning in circles was alarming. It told me that something very wrong was going on with their vestibular sense.

Of all the senses, the vestibular sense is often the most over-looked. Yet it is the most powerful and arguably one of the most essential of our senses. It is also known as our balance sense. There are little hairs inside our inner ear. When we move our body and head in all different directions, the fluid in the inner ear moves back and forth, stimulating these little hairs. This stimulation provides us with awareness of where our body is in space and helps us effectively navigate and move around our environment with ease and control.

Children with a strong vestibular sense will likely have good coordination, accurate body awareness, and skillful balance. For example, they may leap from rock to rock along the ocean with precision and little effort. On the other hand, children with a poorly functioning vestibular sense may consistently run into things, trip a lot, be too close and personal when talking to people, and frequently fall.

Without a strong vestibular sense to provide accurate information about where the body is in relation to its surroundings, all the other senses are also affected, making everything in life more challenging. In fact, the vestibular, auditory, and visual senses are interconnected. If just one of these senses is not working right, the other two are likely to be affected as well.

The late A. Jean Ayres, PhD, a legend in the pediatric occupational therapy world, dedicated her life to researching the integration of our senses—with particular focus on the vestibular sense. She stated, "The vestibular system [network of senses] is the unifying system. All other types of sensation are processed in reference to this basic vestibular information…. When the vestibular system does not function in a consistent and accurate way, the interpretation of other sensations will be inconsistent and inaccurate, and the nervous system will have trouble 'getting started'" (Ayres 2000, 37).

Ironically, due to the lack of efficient movement opportunities today, many children walk around with an underdeveloped vestibular sense. The results: fidgeting, tears of frustration, more falls, aggression, and trouble with attention.

Children develop a strong vestibular sense by having frequent opportunities to move—especially activities that go against gravity. Walking and running offer some vestibular input, but activities that encourage children out of an upright position provide rapid input to the inner ear. In other words, children will benefit immensely by going upside down, spinning, tumbling, and swinging. Most vestibular input can be gained through ordinary play experiences, such as going upside down on the monkey bars, rolling down hills, and dancing until their little hearts are content.

Sight

Sight is an intricate and complex sense. Our sight comes from using light energy to interpret environmental data. It helps us investigate our surroundings and determine our location relative to objects around us. Sight reinforces what children learn through their other senses. For instance, a child may smell muffins cooking in the other room. She then goes over to investigate and confirm with her sight that it was indeed muffins that she smelled.

Sight helps us to determine what an object is as well as some of its properties, such as size, shape, and color. It can also help us recall whether an object is safe to touch, how it feels, and how heavy it is (Roley, Blanche, and Schaaf 2001). Sight is one of the most important senses for survival. For instance, a child with a fully functional visual system will assess how high he is off the ground when standing on a rock. If the rock appears to be too high, the brain interprets the thought of jumping off as a dangerous act. If the rock appears to be somewhat close to the ground, the child may attempt to jump—using his visual memory of how high he successfully jumped the last time.

A newborn is born with the ability to see, but the focus is vague and the baby has trouble differentiating between complex shapes and colors. The first step in developing this sense is to learn to follow a moving object or person with the eyes and then the head. Strong neck muscles support this response of looking and scanning (Ayres 2000). A fully functioning vestibular system supports all six of our eye muscles. It acts similarly to the tripod of a camera, keeping the eyes steady so a child can focus on objects. It also allows for smooth and accurate scanning to search for objects. Therefore, plenty of full-body movement and good neck and eye strength are essential for the normal development and maintenance of the visual system.

Children with visual difficulties may have trouble focusing on faces and other objects. They may have trouble scanning the room to look for something. They may have trouble analyzing the depth of things. Their eyes may be more sensitive to bright lights. This creates safety issues and makes ordinary tasks, such as copying sentences from the board, difficult. Children who need intervention may complain of frequent headaches, rub their eyes often, squint to see, have poor handwriting and drawing skills, have trouble reading, be easily distracted by things in the environment, and have trouble paying attention or concentrating.

If your child has trouble with visual skills, request an evaluation from an optometrist. If lights are too bright and bother your child, whenever possible offer wide-brimmed hats when outside and dim the lights when indoors. Plenty of play opportunities, such as flashlight tag, playing catch, swinging, jumping on a trampoline, and playing on playgrounds, are great ways to get children moving and looking at the same time, helping to organize the sight and vestibular senses together to improve visual skills.

Listening

The auditory sense (interpretation of sound) plays an important role in sensory integration. Listening is a survival-related, primitive reflex that impacts arousal, alertness, and attention (Frick and Young 2012). Listening is a whole-brain, whole-body experience that connects us to our environment and is the precursor to interaction, speaking, reading, and writing. Sounds in the environment, such as birdcalls or other nature noises, give us a sense of the three-dimensional space we occupy. Also, listening to sounds can have an affect on our ability to concentrate and regulate our emotions. Hence, people will listen to music based on their emotions and what they "feel" like listening to. Others will play certain types of music, such as classical, in order to improve their focus when working.

Sounds have many dimensions: intensity (loudness), frequency and pitch (number of sounds per wave), duration (how long the sounds continue), and localization (where sounds are coming from). If your child has trouble processing sounds, she may have difficulty putting all these qualities together (Biel and

51

Peske 2009). Your child may also have trouble tuning out sounds that others can habituate to. For instance, the sound of a fan typically doesn't distract children from doing their homework. However, children who are oversensitive to noise may be distracted by that sound.

People orient to noise as a survival instinct. Noise ignites our postural muscles, which position our body so our eyes and ears are aligned with the object or point of interest. These postural patterns actually foster engagement with the world around us and promote deep breathing and the ability to effectively regulate (control) our senses.

On the other hand, traffic noises and the sounds of sirens or alarms often put people into a fight-or-flight response, leading to short, shallow breathing that sends the body into an alert and upright position. This also brings the focus of the eyes and ears up and out in order to monitor the periphery (Frick and Young 2012). However, our bodies aren't meant to be in a constant state of arousal or stress. In fact, being exposed to loud noises or noise pollution on a frequent basis may actually harm children and their ability to interpret sound over time.

Children may be oversensitive to sounds. Loud noises may really irritate them and cause a fight-or-flight reaction. They may cover their ears and become anxious. Other children may have trouble registering when their name is called. You may repeat, "Johnny… Johnny…" several times before your child even turns to look at you.

The auditory system (network of senses) and vestibular system are housed right beside each other in the inner ear. Therefore, they can significantly influence one another. In fact, moving around stimulates the auditory receptors, and any time you hear a sound you also stimulate the vestibular (gravity) receptors. This is why moving and swinging are often good strategies to use to get children with speech and language needs to start vocalizing more (Biel and Peske 2009).

Taste and Smell

Smell is a primitive sense that alerts us to danger and is connected to our emotions. We smell danger when we come in contact with the smell of smoke, spoiled milk, and rotten meat. When we go to smell something, we sniff the air, and this action creates wind currents to pull odor molecules up to the receptors in the nose. An impulse is then sent up the olfactory (smell) tract that goes directly to the limbic system of our brain, which is our center for emotions, motivation, and pleasure. No other sense plays on our feelings quite like smell (Biel and Peske 2009). For instance, the smell of a favorite meal cooking may conjure up memories of the last time you ate this meal with your sweet grandmother. The smell creates feelings of joy and happiness for you.

The sense of smell is well developed in newborns. It is not further developed and refined in children the way most of the other senses are. Newborns can also taste well. Taste is a sense that tells us something about our environment. Young babies put things in their mouth to gather information about the environment.

The senses of taste and smell work closely together. A perfect example of this is when you have a cold. When your nose is stuffed up, you may complain that you can't taste food. It somehow loses its flavor. That's because we actually rely on our sense of smell to distinguish the majority of foods. We can detect approximately ten thousand different odors. However, we can only taste five things: sweet, salty, bitter, sour, and umami (Biel and Peske 2009). Therefore, many things we think we taste are actually distinguished through our sense of smell.

Children who have difficulty processing the sense of smell and taste may crave or become overly sensitive to particular smells and tastes. They may gag or get nauseated easily. Sometimes when children are intolerant of certain smells and tastes, they start to avoid a lot of different types of foods. Intolerance can create picky eaters. I've worked with children who only eat three foods. Often kids with lots of smell and food aversions tend to pick bland foods, such as bread, plain chips, and pizza. Therapists and parents treat these children by slowly introducing new foods into their diet, one food at a time.

Outdoor experiences, such as gardening, can activate and further develop taste and smell. When children grow their own vegetables, berries, and edible plants, they are more likely to taste them, triggering both the sense of smell and taste. They start sampling different foods that vary widely in texture and taste. The smells of the various flowers and herbs, contrasting with the pungent smell of manure and soil, also expose children to a variety of scents. Things like berry picking, cooking over an open fire, and sampling edible plants found in the wild are also experiences that enhance these senses, and they make for memorable bonding experiences for the whole family.

What Is Sensory Integration?

Sensory integration is simply taking in all of the stimuli detected by the senses (smells, sights, sounds, temperature, balance, gravity) and organizing information about them for functional use. The senses work together to help you effectively process information about your body and the world around you. The more senses that are activated, the more accurate information you have about your environment.

Sensory integration takes all the puzzle pieces and pulls them together in order to create a bigger picture. Imagine climbing a tree in bare feet. You experience the sensation of climbing through your eyes, your feet, your hands, your nose, and even your muscles and joints. All the sensations of climbing the tree come together in one place in your brain. This integration allows your brain to experience the act of climbing as a whole-brain, whole-body experience.

Sensory integration begins in the womb as the fetus feels the movement of the mother. Lots of sensory integration needs to occur in the first year of life in order to get a child to start crawling and later walking. Children continue to integrate their senses as they age through a variety of movement and play experiences. Although all children are born with the capacity for healthy sensory integration, they must develop sensory integration by experiencing many physical challenges during childhood (Ayres 2000).

Difficulties with sensory integration can interfere with many aspects of a child's life. There will be more effort, more difficulty, and more frustration. Children may have trouble knowing where their body is in space and be more prone to injuries. They may have more trouble paying attention in school because they are distracted by their own body or by their surrounding environment. Getting dressed, eating new foods, and doing homework can be a real hassle and cause meltdowns.

In order to integrate the senses, it's important to expose children to a variety of sensory experiences on a regular basis. This involves lots of movement (for example, jumping, spinning, crawling, hopping, dancing), sensory-rich play experiences (for example, making castles at the beach, splashing in puddles, playing in mud), trying new things and new foods (for example, making popcorn over the fire instead of just marshmallows), and even simply listening to the birds chirping. The more exposure your child has to sensory experiences throughout the day, the more integrated and organized the brain, senses, and body become.

As the number of children with sensory issues continues to increase, it is more important than ever that we start to think in terms of prevention. There are not enough occupational therapists to treat every child! Our children will benefit greatly from play experiences that promote movement and challenge the body. Physical challenges will help your child's body to adapt and integrate new senses, helping him or her reach the next developmental level.

THE MIND

The minds of children are rapidly developing every minute. This development is a complex interplay of emotions, interpretation of the senses and movement experiences, memories, planning, and learning. The mind enables your children to be aware of the world around them and to consciously think and make decisions. The mind is both complex and fascinating at the same time. In order to keep things simple, we are going to focus on two primary functions of the mind: social-emotional skills and cognitive skills. These skills are enhanced through outdoor play experiences.

Social-Emotional Skills

Waiting your turn. Following the rules. Dealing with feelings of frustration and anger in a healthy way. Sharing toys. Making new friends. All of these skills describe components of healthy social-emotional development. As with the motor and sensory skills, young children develop social-emotional skills through practice and by making small steps over time.

Infants and toddlers start to develop social-emotional skills early on. You can support this development by simply holding, touching, and speaking to your child and giving him loving attention while letting him play, explore, and follow his interests. Your child will develop new skills when you give him just enough help so that he can be successful without getting overly frustrated. Occupational therapists call this the "just right" challenge. For instance, if your child is trying to climb up a single step onto the next level of flooring, step back and let him—but be there to spot him. Maybe he only needs a little nudge in order to successfully make it to the next level, but even still he did it mostly by himself. The next time he tries, he is likely to have the confidence he needs to advance up the step independently. This teaches him that hard work and persistence are often followed by success.

The same is true with older children. We are there to listen to our children, set clear and consistent expectations, and offer unconditional love and support, but to also celebrate their independence. If your child is ready to ride her bike to a friend's house, let her. This simple act of independence and opportunity to pursue her own interests is likely to boost her confidence. Also, through regular play opportunities with other children, your child learns important negotiation skills, how to take turns, how to put other people's needs before her own, how to comfort others, and many more invaluable social-emotional skills. In fact, playing with other children, away from the adult world, may be one of the most natural and beneficial ways to develop strong social-emotional skills (Gray 2013).

If your child has trouble with social-emotional skills, she may have difficulties playing with other children. She may get easily frustrated and angry and have trouble controlling her rage. Children who struggle with social-emotional skills often don't empathize with the needs of other children. They may have trouble sharing, listening to others, taking turns, and playing by the rules even as they get older and these skills should be in place. It is important that we start cultivating our children's social-emotional skills early. Teach them right from wrong. Listen to them and allow independence whenever possible—especially in an outdoor setting.

Playing independently outdoors with friends further challenges and enhances social-emotional skills. First of all, the natural setting creates a calm, sensory-rich—but not sensory overloaded—environment where kids can play energetically without some of the frustrations, noise, and other stressors that present themselves at indoor play facilities or on school grounds. In nature, away from adults and large groups of peers, children find peace and calm. They have opportunities to work out issues one on one or in small groups. There are no colorful lights blinking, noisy interruptions, or adults constantly checking on them. Time expands as they dive into deep play. Their opportunities for advanced social interactions and problem solving are endless. Therefore, it is imperative that we allow children more opportunities for independent play outdoors. We want them to grow up to be strong, confident, resilient, compassionate, and friendly individuals.

Cognitive Skills

Children also develop cognitive skills through practice and by having plenty of opportunities for play over time. Cognitive skills involve abilities such as paying attention, memory, and thinking. These crucial skills utilize the processing of sensory information to form new memories, to evaluate, to analyze, to make comparisons, and to learn cause and effect. Some cognitive skills are genetic; however, most cognitive skills are learned through real-life situations. In other words, learning and thinking skills can be improved through enriching cognitive experiences.

Children learn best through hands-on and meaningful play experiences. The key word here is "meaningful"—something that is significant or important to the specific individual. When children can make a connection with something that interests them, they are far more likely to engage with all their senses. When their senses are engaged, they are strengthening their sensory skills. And strong sensory integration results in a higher incidence of learning.

To foster learning, be mindful of what interests your child and allow ample time to explore the subjects. For example, if your child is thoroughly excited by the shark exhibit at the local children's museum and wants to spends a good portion of her time there, let her, even if you are bored and would rather see the jellyfish, which is meaningful to you but perhaps not to your child. Sometimes as adults, we feel like we know the best activity to help our children learn something new. However, if we simply step back and follow our children, they will often lead us to what is most interesting and meaningful to them. Just like everyone else, children have specific interests and are naturally curious about the world. They will ask questions, experiment with what they see, try to replicate what they learn in creative ways, and form important neural (brain) connections from their experiences.

When children are deprived of both child-led and play experiences, they may struggle with higher-level thinking skills, such as coming up with their own ideas, problem solving, and other forms of creative expression. It is important that we allow plenty of independent play experiences, in which children have ample time and space to explore, create, and play with friends. It is then, and only then, that they will be able to practice the complex cognitive skills needed for a successful academic career and to reach their intellectual capabilities.

IN A NUTSHELL

There is a common thread that runs through the development of healthy motor, sensory, social-emotional, and cognitive skills. Any time there is a kink in that thread, typically a result of not enough time spent exercising these skills, your child is at risk for a range of problems, from having difficulty making friends to paying attention in school to controlling emotions to even losing the ability to imagine—not to mention being at risk for a range of physical injuries.

The good news is all of these things can be helped and even prevented by giving children ample opportunities for whole-body movement. Due to less time spent developing strength, coordination, and balance, children are becoming more and more unsafe and accident-prone. In order for children to develop any skills of the mind or body, they must practice them daily, ideally through meaningful play experiences. Instead of allowing your children less time to move and play at home and school, try giving them more!

When we constantly say no—"No climbing," "No riding your bike to Henry's house," "No running," "There's no time for that," "Don't touch that," or "Get down from there"—we're likely to see effects on children's development.

We think we know what is best for our children. We are just trying to protect them. However, by constantly rushing children, restricting their movement, and diminishing their time to play, we could be causing more harm than good.

CHAPTER 3

From Restricted Movement to Active Free Play

N ow that we've explored how the body works, it's time to begin answering all the questions posed in chapter one. My answer is quite rudimentary (but important nonetheless): allow your child to have several hours of active free play a day—preferably with other children.

You might be thinking, *Well, my child gets playtime at school* or *My child is involved in sports year-round.* Recess and sports can be great opportunities to get your child moving. However, it's important to realize that children need to challenge their bodies in various ways several hours a day, and to do so with little adult direction. If your child is spending more hours per day in front of a television than outside playing with friends, it's easy to recognize the missed opportunities for playtime: television equals entertained and sedentary children, whereas play equals engaged and moving children entertaining themselves. The math is pretty easy when it comes to screen time. To solve the problem, you might establish a new rule of television as a treat once or twice a week, and add in a compulsory two or more hours of active playtime every day to your child's routine.

But it's a lot harder to see missed opportunities in other aspects of your child's life. Consider that long, uninterrupted hours in the classroom, at the kitchen table doing homework, and in a car seat on long commutes also steal quality playtime from children. Not only that, the physical restraints caused by sitting can be detrimental in many other ways—physically, mentally, and emotionally—as I've touched upon.

This chapter explores the serious effects of putting too many limits on movement. We'll start by looking at the most common movement restrictions that kids face today. Then I'll explain what "active free play" really means, what it entails, why it's important, and how you can foster it.

THE EFFECTS OF DAILY RESTRICTIONS ON MOVEMENT

As a parent of two young children, I can empathize with the parental fear that habitually gets in the way of childhood risk taking. Our instincts often take over, and we shout "Be careful!" or "Slow down!" as we watch a child climb on top of boulders or speed across uneven terrain. However, as a pediatric occupational therapist who spends countless hours observing children playing in a natural environment, I also know that restricting children's movement and limiting their ability to play outdoors can cause more harm than good.

Children's ability to move and play are being restricted more than ever before. At a young age, children are placed in baby devices that restrict natural movement and good posture, leading to altered walking patterns and unnecessary full-body weakness. Older kids are required to sit for hours on end during school. After school, they are overscheduled with organized sports, art classes, music lessons, and more. This leaves little time for children to be involved in active, spontaneous free play—the type of play that stimulates the senses and gets the heart pumping.

Let's take a closer look at how each of these restrictions can impact our children.

Beware of the Baby Devices!

Baby devices are hard to resist. They allow you to take a shower, clean the house, and finish the taxes—while keeping baby safe from injury. However, if babies spend too much time in these devices, they can actually cause physical complications and possibly developmental delays, such as being late to sit, crawl, walk, and so on (Crawford 2013).

The rise in baby-device usage is partly to blame on the "back to sleep" campaign issued by the National Institute of Child Health and Human Development in 1994. This campaign was created in an effort to decrease the amount of sudden infant death syndrome (SIDS) cases. SIDS is the sudden death of an infant less than one year of age. Many experts felt that a possible cause of SIDS was putting infants on their stomach to sleep with soft materials, which could cause suffocation.

After this campaign began, there was a significant decrease in the number of children who died from SIDS. However, many doctors and developmental therapists (for example, occupational, physical, and speech therapists) feel that this campaign triggered an unintentional rise in the use of infant devices to contain children for most of their day. Doctors and therapists were seeing a simultaneous rise in children with flattened heads, developmental delays, coordination difficulties, muscle weakness and abnormalities, and altered walking and movement patterns. It got so bad that therapists coined the term "container baby syndrome" (CBS) to describe what they were seeing.

Researchers say the number of CBS diagnoses increased 600 percent between 1992 to 2008, culminating with one in seven children being diagnosed with CBS in 2008 (American Physical

Therapy Association 2008). At Children's Healthcare of Atlanta, senior physical therapist Colleen Coulter-O'Berry says that a decade ago her clinic saw about fifty children annually who needed therapeutic helmets to correct head shape. Now the clinic sees about five hundred kids a year who need a helmet (Manier 2008). Evidence is starting to mount that infants who spend a lot of time on their backs are also more likely to have slight delays in motor skills. A 2006 Canadian study published in the *Journal of Pediatrics* found that 22 percent of babies who slept on their backs had some delays in motor skills such as sitting up, rolling over, and climbing stairs (Manier 2008).

Although baby equipment such as infant carriers, backpacks, and bouncy chairs can be helpful and convenient, using them to contain infants for most of the day immobilizes them and places continuous gravitational forces on certain parts of the body. Over time, this can alter walking and movement patterns.

Constantly being on the back also allows for little to no movement of the neck or spine. Babies need movement through the center of their body in order to start developing essential core and neck strength. A strong neck and core lay the foundation for many other developing skills, such as fine motor skills, visual skills, body awareness, coordination, and balance. Frequent time spent on the floor (both on the belly and back) allows babies to move their limbs freely, explore and touch their surrounding environment, and start to develop strong muscles and bones. Chapter eight dives deeper into this discussion of movement and tackles the sensory and motor benefits of getting little ones outdoors on a daily basis.

The "Sit Still" Mandate

A local elementary-school teacher informed me that children are expected to sit for longer periods of time than in years past.

Maybe you've noticed this trend with your own children and are surprised at how much time the school system expects them to sit. This increase is likely due to the expectations of teachers to fit in more and more curriculum at an earlier age. In fact, even kindergarteners are expected to sit for thirty minutes at a time at many schools.

One kindergarten teacher tells me that she feels like she is under constant pressure to produce "results" in her students. By the end of kindergarten, children are expected to read, write, add, and subtract; if they don't learn these skills, the children have failed, along with the teacher. She went on to tell me that in the United States, many people are pushing for teachers' pay to be affected by their students' test scores, which is why many teachers are getting so serious with their push for more academics.

A preschool teacher tells me that even she feels compelled to push children at a young age. On top of that, teachers feel so much pressure to document and justify what they do and why they do it that the relaxed, playful environment is often compromised. As academic demands on children increase, many children are asked to take a seat. They aren't sitting for just a brief period, followed by lots of rich opportunities to learn through hands-on experiences. On the contrary, the majority of children are expected to sit for *hours* every day. This lack of movement combined with an unrelenting sitting routine is wreaking havoc on children's minds and bodies.

I recently found out about a local middle school that got rid of its recess in order to fit in more curriculum. Curious, I decided to experience their school environment firsthand. I sat in one of the classrooms as if I were a student myself. Except for the brief periods of getting up and switching classrooms, we sat for almost *three* hours straight. At one point, I looked down at my leg and noticed it was bouncing. I was starting to fidget. I looked around and noted that my fellow classmates weren't much better off.

Those who weren't fidgeting were slumped over their desks or slouched way back in their chairs.

I started contorting my body into awkward positions to keep from daydreaming. It was useless; about forty-five minutes into the class nothing the teacher said was registering anymore. Children started raising their hands to go use the bathroom or to sharpen pencils, anything to get up and get moving. I was planning on staying the whole day. I just couldn't do it. I decided to leave shortly after lunch. Sitting for the three-hour morning had completely exhausted my mental energy, and I craved an afternoon siesta.

How can we expect young children to sit for hours at a time when it's hard to do so as an adult?

The bottom line is this: it is hard to pay attention when you are not allowed to move for extended periods of time. This is one reason *why* children fidget! In order to remain alert, children activate the vestibular system by moving back and forth in their chairs. This movement turns the brain on to pay attention. It's not that these young students are trying to be disruptive or are not interested in learning—on the contrary. They are in fact straining all of their resources in order to listen and learn. The classic sign that kids are not getting enough movement throughout the day is when they start wiggling, rocking, and twisting their bodies about.

Not wanting the children to get hurt or to distract others, the teacher asks them to sit still and pay attention. This is a dilemma for the children. If they keep their body and head still, it reduces activation of the brain, making it harder to do what the teacher wants them to do—pay attention and learn. But in order for children to learn, they must be able to pay attention. And in order to pay attention, children need to move.

Sitting for practically the whole day is also unnatural and can be harmful to our bodies. We were made to move, not to remain sedentary. When we sit for long periods of time, day after day, our bodies start to succumb to these unnatural positions and a sedentary lifestyle. This can cause atrophy of the muscles, tight ligaments (where there shouldn't be tightness), and underdeveloped senses—setting children up for weak bodies, poor posture, and inefficient sensory processing of the world around them.

Screen Time Is Taking Over

According to the American Academy of Pediatrics (2013), a recent study states that the average child spends *eight* hours a day in front of screens (television, video games, computers, smartphones, and so on). Older children and adolescents are spending an average of eleven hours a day in front of screens. And 75 percent of twelve-to-seventeen-year-olds have their own cell phones. Nearly all teens participate in text messaging.

Once at a TimberNook camp, I witnessed a young child go up to a tree and ask where the buttons were. I've seen other children try to swipe a toy to get it to engage and entertain them. Lots of video games and television shows are designed specifically for entertainment purposes. Many parents feel like they need to constantly entertain their children every minute of every day. When they need to do chores or take a shower, they use the television or video games as a babysitter. At a really young age, children become accustomed to this mode of being. The effects? Children become unable to think for themselves and lose the ability to imagine, and their essential play skills are hindered.

One eight-year-old girl who came to camp a few summers ago had the hardest time during free play. She would come up to the staff members and ask, "What's next on the schedule?" "It's time to play now," we would respond. She would then find a seat on a stump. Incredulous of this notion of free play, she refused to leave the stump unless she needed to go to the bathroom or it was time for a more structured activity. No matter how many times children invited her into their play, she refused to move.

A few college professors state that this lack of independence and creativity is even appearing at the college level. They are starting to see more students who have trouble with simple problem solving, creative thinking, and even answering essay questions. One professor told me that we need more game *designers*, not more game players.

Most importantly, video games and watching television are addicting and take precious time away from play—especially play outdoors. Instead of being fully immersed in a game of baseball out in the field with friends, where children practice negotiating the rules, running the bases, and stimulating the senses in a healthy way, children are indoors and sitting—again. Sometimes parents put children in front of the television with the hope that it will help their children relax or calm down; however, usually the reverse is true.

Bright colors flash before them, stimulating the fight-or-flight response in the brain—but with no release. This lack of release is why, when you turn the television or video games off, children are often angry. Their brains were stimulated without the chance to move and react. There is nothing wrong with an occasional video-game match or movie. However, such screen time really should be considered an occasional treat, something that doesn't become habit or that children feel entitled to have every day. Their time is so precious already. Instead, consider giving them real, authentic play experiences. Try saving the screen time for a rainy day or special occasion.

Overscheduled and Overwhelmed

You probably have a good handle on what you can fit into a typical day and week. As a busy parent, you've likely got work, errands, household management, and child-rearing pretty well scheduled so as not to get too stressed out. Sure, some days are crazier than others, but you are probably skilled at knowing your limits, when to say no, how to prioritize, and how to wind down. But have you ever stopped to think about what your day is like for your child, whose part of the brain responsible for higher-level reasoning (for example, the ability to make decisions, understand consequences, or prioritize) hasn't fully developed? Let's look at the day of a typical modern kid.

Sarah, a sweet nine-year-old, gets up and dresses quickly in the morning, keeping in mind that if she eats fast enough her mother will let her watch cartoons before leaving for school. As soon as she finishes her last bite, she watches reruns of "Looney Tunes." Twenty minutes go by in a flash.

Sarah lives in the country, so it takes about twenty-five minutes to drive to school. Her mom feels guilty about the long drive, so she lets Sarah play on the iPad until they arrive. "Everyone, please take your seats!" the teacher shouts as Sarah approaches her desk, where she sits most of the day, except for a brief snack, quick lunch, and a twenty-minute recess. Then it is time to go home.

After another twenty-five-minute commute home, Sarah is feeling energetic after a long day of sitting. She instantly heads toward the swing set in the backyard. "Not yet," Mom catches her. "Homework first." Sarah groans, shuffles to the dining-room table, and pulls out her assignment.

"Argh…" Sarah is literally trying to pull her hair out. "I hate this! I can't *do* this!" It takes her about ninety minutes to complete the homework on a good day. By the end, she is exhausted. After two bouts of crying, she feels angry and spent. "Can I play on the iPad for a little bit?" She asks her mother. Her mother, thinking Sarah has definitely earned it this time, says, "Sure. But remember, we need to leave for Girl Scouts in thirty minutes."

After Girl Scouts, Sarah and her family use the drive-through on their way home since it is already a late night. When they get home, Sarah grabs her Harry Potter book and reads for thirty minutes before it is time for lights out. Tomorrow, she'll do this routine all over again. Only instead of Girl Scouts, she has basketball practice.

Does this routine sound familiar? Maybe there are a few variations in your family. Maybe you cook dinner in a Crock-Pot so it's ready to eat when you get home. Maybe you have more than one child, and this scenario seems like nothing compared to your schedule. Regardless, many of our schedules have become busy— *very* busy. This leaves children little time for free play outdoors— the type of play that rebalances them and gives them respite away from an unnecessarily demanding world.

Parents today presume that playing a team sport is superior to playing freely at the park. Don't get me wrong, I believe sports can offer great value: they teach children responsibility, team ethics, perseverance, patience, stamina, endurance, and the challenge of competition. Problems arise when this belief causes parents to replace active free play with sports, leaving no time for children to engage in imaginative, child-driven, and sensory-balanced play.

Keep in mind that organized sports have changed in the last thirty years. In the early 1980s, I remember playing softball and soccer. Back then, we practiced about once a week and had the occasional game on a Saturday. Most of my weekends and weekdays were still filled with hours of outdoor play with my friend Jessica, biking around town, going to garage sales to buy items for picnics in the park, and trying to earn cash by washing our neighbors' cars.

Now it appears that organized sports have all but taken over the extracurricular lives of children. Studies show that 60 percent of boys and 47 percent of girls in the United States are on a sports team by the age of six (Kelley and Carchia 2013). Even children as young as three and four are suiting up and getting out there to practice for a team. Equipment is being altered, helmets fitted, expensive uniforms purchased, and private lessons bought to give young children an "edge." Practices and games are no longer just once-a-week fun and laid-back sessions. They can number up to three or four a week for elementary-aged children—an intensity that was once saved for middle-school children who are mentally and physically better equipped for this kind of demanding schedule.

Not only are organized sports becoming more of a commitment, but children are often playing more than one sport at a time, being driven to sibling's activities and games, taking lessons, and joining clubs. Why have sports become so intense? Is this an attempt to keep kids busy? What about letting kids entertain themselves? What about teaching children about balance in life? With organized sports becoming a given, rather than an option, in the lives of our children, we've lost sight of an important principle that is crucial to healthy development: organized sports can be an okay way for kids to get exercise, but they should be a supplement to active free play. Sports should be the icing on the cake, *not* the cake, when it comes to providing an environment for kids to thrive developmentally.

When children develop and organize their own athletic games outdoors and without adult interference, their experience of playing sports is enriched on many different levels:

- An impromptu game is always a choice, not an obligation.

- It's seen as a form of play.

- Children will naturally create their own rules and set their own boundaries.

- Kids determine when to stop on their own—when they feel done.

- Children learn team ethics (for example, they create their own rules and work toward common goals).

- Children experience competition (for example, they learn that there is sometimes a victor and experience failure, which is necessary in order to develop the skills of persistence, control, and hard work).

- Children learn how to be empathetic and accommodate other children's needs—not just their own.

- Children feel less pressure and anxiety when they take some ownership of the rules and regulations.

- Children have a greater sense of accomplishment because they helped design the play scheme or game.

- Children regulate their own physical abilities (for example, they determine when and if they want to be the pitcher or the goalie, when they need to sit out and rest, and so on).

- Everyone gets a turn—no more waiting on a bench.

I once treated a child who was very anxious. When I asked his parents what his day looked like, they told me he had one or two extracurricular activities *every* day of the week, including the weekend. They barely had time for their occupational therapy sessions—never mind finding time for him to simply play.

With long hours of sitting in school followed by inappropriate amounts of homework and being rushed to after-school activities, it is no wonder our children are having increased anxiety, difficulty playing independently and creatively, and trouble developing sensory skills.

Through child-initiated play, children naturally develop strong muscles and sensory systems, learn creativity, and develop healthy social and emotional skills. However, we need to allow them time to do this. When their schedule is booked full with structured activities, there is little time for active free play outdoors—the type of play that gets children thinking, moving, and creating, using both their mind and body in ways that following adult direction will never match.

ACTIVE FREE PLAY

Aw…to be free. *Free* to play whatever you want. *Free* to explore. *Free* to wander away from the house. *Free* to make mistakes. *Free* to jump, spin, dance, shout, and climb. *Free* to take risks. This is *active free play*—moving the body, stimulating the senses, and igniting the imagination so that the whole body and brain are engaged at once. Neurons are firing on all cylinders as children explore their surroundings. They are fully *alive*.

Active free play outdoors is becoming rare, a thing of the past. Yet it is more important than ever that we stop overscheduling our children and start to reintroduce more opportunities for play and movement. Children's minds and bodies depend and thrive on active free play.

Give the Gift of Free Play to Your Child

Unrestricted and unsupervised play is one of the most valuable educational opportunities that we can offer our children. I once had the privilege of hearing Peter Gray speak. He is what I would call a "play expert"—a scientist and researcher who studies the evolution and theory of play. He defines "play" by talking about its distinct qualities. First, he says that play is self-driven and self-directed. You always have a choice whether you want to play or not. He says, "The ultimate freedom in play is the freedom to quit" (Gray 2013, 141).

According to Gray, when adults take over and direct play for children, it is no longer considered play. For instance, adult-led academic games may be fun for kids who choose to play them in school; however, they may feel like punishment for kids who didn't make that choice (Gray 2013). A game of "kick the can" or pickup basketball that kids play on their own is "play." A little league game directed by adults is not.

Play is guided by mental rules. As Gray states, there is no such thing as "unstructured play," because when children get together, they form their own rules (2013). For instance, if children play "house," they are likely to assign roles to each other. "I will be the mother," one child shouts out. "No, how about you be the sister?" another child suggests. "Okay, only if next time I get to be the mother," the first child answers. Their games can become quite complex and structured, whether adults see them that way or not.

Oftentimes in the woods at TimberNook, children create societies and even their own hierarchies. They take on roles such as "top spy," "teepee protector," and "first commander" and report back to the leader who sits perfectly perched in the tree donning a feathered mask. They go against other teams and decide on methods of "attack" and strategies for how to hide their "goods." Sometimes these intricate games are remembered and picked right back up the following summer. These play ideas are never suggested by an adult. In fact, the adults watch from a distance with awe as the elaborate play unfolds in front of them.

Play is also imaginative. It is serious for the children, but also not serious. It feels real to them, but it isn't real (Gray 2013). Children will often play so deeply that they appear to be in another world. I once witnessed a little boy stop and ask another boy, who was involved in superhero play, "Wait! Is this real?" Sometimes they take breaks from their pretend play to eat a snack or lunch or go to the bathroom. "Freeze!" one child yells. "Snack time." Another child involved in the play then seeks reassurance that this is in fact just a break from their play. "We are going to play after, though, right?"

Play is motivated by means more than ends. When we engage in an activity purely to end it, that is not play. For instance, reading a book in order to do well on a test is not play. Most often, children don't want their bouts of play with friends to end. Play is constantly evolving and developing into new forms of play and play schemes. Play is also not passive but requires constant assessment and engaged minds (Gray 2013). Yet it shouldn't be stressful to children. They may not always be happy during play experiences, as rules get negotiated and feelings get hurt from time to time. However, they always have the choice to not play and the freedom to change the play experience.

Giving your children time to engage in free play is like giving them a very special gift—a gift that keeps on giving, preparing children for adulthood by cultivating and nurturing essential life skills. Play allows children opportunities to get creative, to practice regulating emotions, to enhance social development, and even to learn about themselves in the process.

Having the ability to play away from the adult world opens up many opportunities and feelings of freedom. It is fertile ground, a blank slate on which children develop their own stories and preferences. Children take ownership over their play experiences and get creative with what is around them. A stick can become a wand, a weapon, a fishing pole, a horse on which to gallop, a building material, or a tool. Leaves can become an ingredient for soup, a prop for art, medicine, money, decor, and so on. The possibilities are endless.

When offered their own free play, children decide what they want to play and with whom. Coming up with play schemes or ideas is not only great fun but also a mental challenge for children. If they want to play with others, they have to learn how to invite others to join their play. They also learn how to present their play opportunity so it sounds worthy and pleasurable. Once there are a few children involved, they start negotiating their play schemes and creating a more elaborate form of pretend play.

This delicate back and forth teaches them how to compromise and work with others as well as how to self-direct and generate creative ideas. These are important characteristics that foster creativity, independence, and interpersonal intelligence (the ability to relate to and understand other people). Interpersonal intelligence really needs to be learned through real-life experiences; it can't be taught through textbooks or lectures. Role-playing, which children who struggle with talking and interacting with others practice in social-skill groups after school, won't be as effective as practicing and experiencing firsthand how to make and keep a friend.

Through play and risk taking, children also learn about themselves. They learn their interests, their abilities, and how to regulate their emotions. Children learn to work through frustrations, fear, and anxiety as they successfully climb onto large rocks to have a picnic with friends or by having to negotiate a new play scheme because a friend just threatened to not play if they didn't.

Children test their limits both physically and mentally, growing stronger each time they play. They develop a sense of confidence as they climb a tree a little higher or another child agrees to play with them when they ask. They learn patience and how to persevere in order to keep the game going. Through free play, children become flexible, resilient, and capable. Free play lays the foundation for a successful working career and the development of long-term relationships as an adult.

The Right Kind of Movement

In an effort to fight the nation's rising obesity problem, in the past decade there has been a significant increase in the number of available exercise classes, youth sports teams, running programs, and fitness camps for young children. According to the US Youth Soccer Organization, the number of children who participated in youth sports between 1981 to 1997 grew by 27 percent—and continues to grow (Kelley and Carchia 2013). Everything from yoga classes to enticing running programs are being implemented all over the United States. However, the waistlines of America's children continue to grow, and obesity is still on the rise. According to the Centers for Disease Control (CDC), 16.9 percent of children were obese in 2009. This is triple the number of children who were obese in 1980. Obese children often grow up to be obese adults. By 2030, the CDC predicts that 42 percent of all Americans will be obese (Ogden et al. 2012).

Clearly all this time that adults are spending organizing exercise opportunities for children and encouraging them to play multiple sports, even at the age of three or four, hasn't forestalled the epidemic. The well-meaning push for organized sports has overlooked the fact that for generations children simply played outdoors and didn't don fancy jerseys, nor were they subjected to rigid schedules and rules. They got their exercise by playing pickup basketball with friends after school, by playing tag out in a field for hours, or by making forts in the woods with no adult present.

Not only did this form of free play ignite their imaginations and creativity, it also challenged their bodies in a multitude of different ways for hours on end. No one determined the rules for them; they set the rules, and then through determination and interest they tested their own abilities. They challenged their senses and their muscles and learned to persevere and overcome personal goals and obstacles.

A good example of active free play is good old-fashioned pond hockey. Even though people still play pond hockey today, it is now rare for children to take off on their own to start up their own games on an actual pond or lake. In the past, children, both young and old, would have walked to the pond by themselves with skates in hand. If there were enough kids to play and the ice was thick enough, they'd set up the goals, which could be anything from an egg crate to ski poles stuck in the ice.

Children would pick their own teams based on their abilities and set their own rules. They'd have to regulate how hard to hit the puck in order to successfully complete passes to young kids versus older and more skilled kids. The bumpy ice further challenged the children's motor and balance skills as they maneuvered the uneven terrain.

Most importantly, the children were out there by choice and to have fun. No adults were standing around, shouting orders. In the name of fun, children tolerated and endured skating for hours, sometimes even until the sun went down. They challenged their bodies and their limits, while enhancing their skating skills every time they went back on the ice. Active free play outdoors develops the senses, builds strong muscles and bones, and fosters a healthy immune system.

Active Play Builds Strong Muscles and Bones

Active free play develops the strong muscles and bones needed for stability, injury prevention, endurance, and strength. This development is further enhanced when the play takes place outdoors. Studies of children in Norway and Sweden compared preschool children who played on relatively flat playgrounds with children who played among large rocks, trees, and uneven terrain.

They found that children who played in natural areas tested better on their motor skills, especially in regards to balance and agility (Grahn et al. 1997).

Katy Bowman, a biomechanical expert says that when children are exposed to low and gentle forces multiple times each day, even as babies, they quickly develop the muscle strength they need to support their own weight (Crawford 2013). This happens naturally through play. Children pick up heavy sticks to use for building, run up and down a slanted beach over and over again to refill a bucket of water, and climb over fences and fallen logs to get to the other side of a field. When children play outdoors, they are naturally motivated to move—strengthening their muscles with each move, each step, and every encounter with nature.

The outdoor environment is unpredictable. Rocks, sticks, and logs vary in size and weight. Children must learn how to regulate their strength to pick them up, each time testing their physical limits. Climbing, hanging, and digging also help develop the muscles of the core and upper body. Navigating the uneven and varying terrain that the outdoors offers (especially with bare feet) challenges the muscles of the legs, ankles, and arches of the feet. Developing muscle strength through play helps provide stability and strength for the spine and limbs.

Tendons and ligaments are also strengthened through active play. When children move and play outdoors, they are naturally stretching and lengthening their connective tissue and increasing their range of motion. For instance, when they reach up high to grab a tree limb or pull themselves up onto a large boulder, they are improving their range of motion. On the other hand, when connective tissue remains in a loose state consistent with nonuse, it will gradually shorten and become tight. Tight ligaments, tendons, and muscles are more prone to tears. Children need to move and play to keep connective tissue flexible and healthy in order to prevent injury.

Just like muscles, bones develop and strengthen through varying the type and amount of gravitational loads to the body. Things like running on uneven terrain, jumping off of small rocks, and stomping in puddles are all great ways to provide healthy weight-bearing opportunities to strengthen bones. When children don't experience enough of these movement opportunities, their bones break down and a release calcium, which is reabsorbed by the body, leaving bones more brittle and weak and increasing the risk of fractures (National Space Biomedical Research Institute n. d.).

Dr. Sheref Unal, assistant professor of pediatrics at the Southern Illinois University School of Medicine, emphasizes the importance of developing strong bones from the moment a child can move around. He states that weak bones in children are caused by sedentary lifestyles and lack of exposure to vitamin D from sunlight. Dr. Unal reports, "If they don't achieve good bone strength during childhood, it can lead to things like osteoporosis or brittle bones later in life." He recommends that children go outdoors for vitamin D and experience active play in order to develop strong and healthy bones (Southern Illinois University School of Medicine 2007).

The Benefits of Heavy Work

Active free play outdoors offers many opportunities to stimulate the senses of the muscles and joints, developing a strong proprioceptive sense. The senses of push and pull that happen when children interact with their environment (for example, picking up heavy sticks to build a fort) create new gravitational loads and adaptations that strengthen bones and muscular tissue over time, offering children increased awareness of their muscle's capabilities and of their positioning for better body awareness.

Great examples of "heavy work" outdoors include pulling a sled up an inclined hill, digging in the dirt to plant new flowers, and climbing a tree. All of them apply added forces and make the muscles work harder, providing excellent sensory input to the muscles and joints. For instance, walking up an inclined hill adds gravitational loads and expectations to your core and leg muscles, essentially making them work more. Digging in the dirt increases sensory stimulation to the muscles and joints surrounding the shoulder complex, arms, hands, and wrists. Climbing a tree provides increased awareness to the many muscles and joints that are activated while scaling upward.

The more time children play outdoors, the more they are exposed to natural forms of heavy work. Over time their body adapts to the various loads and forces to develop better body awareness and a sense of the right amount of force to apply when interacting with their environment. In other words, they learn to regulate how much pressure to apply when playing a game of tag or holding a baby chick.

Having a strong proprioceptive sense due to years of outdoor play comes in handy when children later learn how to do more precise work, such as sawing the limbs off a dead tree or using a sewing needle with great skill and accuracy. Therefore, it is essential to offer your children plenty of outdoor time to equip them with a good sense of body awareness and to learn how to accurately assess and engage with the world around them.

The Benefits of Spinning

Children will naturally spin, go upside down, roll down hills, and move their body in all different directions when given the opportunity. Have you ever observed your child spinning just for fun? As children play and move through space, they activate the hair cells in the inner ear. This activation sends motor messages throughout the spinal cord, contributing to the maintenance of muscle tone and body posture (Ayres 2000). In essence, spinning and similar movements contribute to the healthy development of the vestibular sense. As you learned in chapter two, the vestibular sense lays the foundation for many of the other senses. The vestibular sense is necessary for attention, balance, eye control, postural strength, and more.

Spinning in circles is one of the best activities to help children gain a good sense of body awareness. It basically establishes their center, or core. Until children have good awareness of where their center is, they will have trouble establishing a dominant side for writing and throwing, and coordinating the two sides of their body will be difficult. This is why it is important to allow your child to roll down hills and spin in circles just for fun.

David Clarke at the Ohio State University College of Medicine has confirmed the positive results of spinning. With fewer opportunities to spin and the disappearance of merry-go-rounds (you'll learn more about changes in playgrounds in chapter six) comes a new worry: an increase in learning disabilities. Clarke's studies suggest that certain spinning activities lead to alertness, attention, and a sense of calm in the classroom (Jensen 1998).

I've heard from many children that they are not allowed to spin on their swings at school or even to just spin in circles for fun—that this has become a forbidden activity in recent years. Adults have told me that they are nervous that their children might get dizzy, fall, and hurt themselves. However, spinning actually helps them develop better body awareness and improved attention. Over time, with many opportunities to spin, go upside down, and stimulate the vestibular sense, children develop the ability to navigate their surroundings with ease, strength, and accuracy. They become more coordinated, sure-footed, and less likely to trip or run into things. The vestibular sense also improves their ability to concentrate in the classroom.

Therefore, we should be careful about eliminating play opportunities that foster healthy sensory and motor development. If we simply set up an environment that allows for full movement experiences and put fewer restrictions on children, they will naturally seek out the opportunities they need for healthy sensory integration on their own.

Strengthening the Immune System

The role of movement should not be overlooked when addressing the increase in colds, illnesses, and allergies in childhood that we learned about in chapter one. When children move on a regular basis, they increase blood flow to different parts of the body, increase oxygen intake, and activate their lymphatic system. The lymphatic system transports *lymph*, essentially a clear and colorless fluid that contains essential infection-fighting white bloods cells, to different parts of the body. This system helps rid the body of toxins, waste, and other unwanted materials.

The lymphatic system is vital for maintaining a healthy immune system. However, unlike the circulatory system, it doesn't have a pump and moves only in one direction. This means it relies on the movement of our muscles and our diaphragm (muscle that aids deep breathing) in order to effectively replenish the system and get rid of toxins. If a lymph system becomes less active due to lack of movement, the body can be less protected against colds and illnesses.

Activities such as jumping and other vigorous up-and-down movements are reported to increase lymph flow by fifteen to thirty times. Dave Scrivens, a certified lymphologist, says that the "lymphatic system is the metabolic garbage can of the body. It rids you of toxins such as dead and cancerous cells, nitrogenous wastes, infectious viruses, heavy metals, and other assorted junk cast off by the cells. The movement performed in rebounding provides the stimulus for a free-flowing system that drains away these potential poisons" (2008). Movement also helps stimulate the gut, which aids good digestion and normal bowel movements. Lastly, increased heart-pumping exercise has been known to expand the lungs over time, increasing the intake of oxygen.

In order to increase our children's endurance, strength, and tolerance for playing outside so they can reap countless health benefits, we need to expose them to the great outdoors on a frequent and regular basis. The *Physical Activity Guidelines for Americans*, issued by the US Department of Health and Human Services, recommends that children receive sixty minutes or more of physical activity each day. In my interview with Dr. Faria, a well-respected chiropractor, she stated that "This is just to prevent disease. Sixty minutes of movement a day is not enough to promote health in children."

Children need to move for *hours* every day in order to reap the sensory, cognitive, and health benefits that develop strong and capable children. Just like exercise for adults once a week isn't enough to make adequate changes to our fitness level, exercise once or twice a week at soccer practice is not enough sensory input to make lasting changes to a child's sensory system. Children need to get outdoors, stimulate the senses, and move their body in all different directions on a daily basis. We were made to move, adapt, and move some more.

HOW MUCH ACTIVE PLAY IS ENOUGH?

How do you know if your child is getting enough active play? Children should be getting daily movement experiences throughout the day in order to develop strong and healthy musculoskeletal and sensory systems. This lays the groundwork for the development of higher-level mental and physical skills as children age. Ideally, kids of all ages should get at *least* three hours of free play outdoors a day.

Infants (one month to twelve months)—Infants benefit from having opportunities throughout the day to be active and outdoors. Physical activity encourages organization of the sensory system and important motor development.

Toddlers (twelve months to three years)—Toddlers could benefit from at least five to eight hours worth of active play a day, preferably outdoors. They will naturally be active throughout the day. As long as you provide plenty of time for free play, they will seek out the movement experiences they need in order to develop.

Preschoolers (three years to five years)—Preschoolers could also use five to eight hours of activity and play outdoors every day. Preschoolers learn about life, practice being an adult, and gain important sensory and movement experiences through active play. It's a good idea to provide them plenty of time for this.

School age (five years to thirteen years)—Young children up to preadolescence could benefit from at least four to five hours of physical activity and outdoor play daily. Children in elementary school need movement throughout the day in order to stay engaged and to learn in traditional school environments. They should have frequent breaks to move their body before, during, and after school hours.

Adolescents (thirteen years to nineteen years)—Adolescents could benefit from physical activity three to four hours a day. Children in the teens still need to move in order to promote healthy brain and body development, regulate new emotions, and experience important social opportunities with friends out in nature.

SPECIFIC TIPS FOR FOSTERING STRONG AND CAPABLE KIDS

Not only is allowing children enough time to play important, but the quality of that time will determine the amount of developmental benefits they receive. The following are some basic tips on how to foster strong, healthy, and capable kids. These tips will be explained in greater detail in the following chapters.

- Allow adequate time *every* day for children to play outdoors.

- Take frequent movement breaks throughout the day in classroom settings.

- Give children adequate time to play at recess. (This will be addressed in chapter seven.)

- Allow children to move prior to going to school, such as helping with chores outside.

- Let them play outdoors when they get home from school for at least a few hours.

- Younger children don't need to do organized sports or activities; they'll get adequate exercise simply through play.

- Invite children to come over and play with your children outdoors for the day. Your children are likely to be more independent in their play with friends around.

- If you live in a neighborhood with other children, let your children go and play with friends.

- Let children take risks—even the youngest ones—such as jumping off a small rock or walking on the side of a curb.

- Instead of entertaining your children primarily through adult-led activities, inspire movement by using the environment (set up a rope swing outdoors, provide a bike and a basket, put a wagon outdoors). Let them take the lead on what they'd like to do.

- Most importantly, give your children the gift of time to move and play every day!

IN A NUTSHELL

You don't need to structure your children's activity during recess or when they are home. Simply step back and allow them ample time to move and play outdoors on their own. Your children will naturally create their own play opportunities and seek out the type and amount of movement they need—without the need for adult intervention. Active free play is critical for developing healthy bodies and minds. It allows children to develop creativity, independent thinking skills, confidence, emotion regulation skills, strength, and healthy sensory and immune systems.

CHAPTER 4

The Therapeutic Value of Outdoor Play

We've learned how important active free play is to developing minds and bodies—especially when it takes place outdoors. Now we are going to dive a little deeper into what specifically makes the outdoors *therapeutic*.

You may be thinking, *My kids get a ton of free play and are moving about all the time! Why does it have to take place outdoors? What is so special about nature that humans can't try to replicate this experience? How is rolling down grassy hills any different than rolling down a ramp at a gymnasium? Isn't getting messy with shaving cream in the bathtub just as good as getting messy with mud outside?* I will answer these questions in this chapter.

WHY OUTDOORS?

Virtually everything that can be done indoors can also be done outdoors—and not just traditional play activities and games. With some creativity and a little forethought, even ordinary tasks such as eating and bathing can actually happen outside for a fun, enriching, and memorable experience. So grab the picnic supplies and that industrial-size bucket!

Let's consider the girl who grabs her magic wand from her toy box and twirls around her room, turning her books and dolls into frogs and princesses. She imagines a bad witch coming through her window, so she considers building a fort out of bed sheets and pillows. But she's worried she'll get in trouble for messing up her room, so she changes her mind and decides to play "dress up." She wants to be a ballerina, but all she can find is her fairy dress, which won't do as a substitute. She decides to do something else.

But that same child outdoors plays quite differently. She finds a rough, crooked stick that instantly becomes a magic wand. A large hill becomes a place to escape an unwelcome dragon, and she races up it. But the wind in her face becomes a massive storm, so she rolls down the hill, where leaves form an imaginary pool of lava that she must carefully traverse by hopping from rock to rock.

Indoors, there are rules to follow. And objects have a specific purpose. Even toys that are meant to inspire creative play can be seen as having a single functional role, leaving kids feeling limited by the very items that are supposed to bring hours of play. The outdoors, however, has fewer rules and guidelines. And objects in nature, because they don't seem to have any inherent function or usefulness, actually inspire kids to use their imaginations, challenge their thinking, and test their physical limits—far more so than almost anything made in a factory. True joy, a sense of play, and confidence overcome children who play outdoors.

When I advocate for children playing outdoors, I remind parents of three key factors that I've never seen successfully duplicated in any indoor environment:

- The outdoors offers a perfectly balanced sensory experience.

- The outdoors inspires the mind.

- The outdoors is an ideal setting for evaluating risks and accepting challenges.

Now let's look at these factors in more detail.

The Outdoors Offers a Perfectly Balanced Sensory Experience

Imagine your child walking barefoot through a meadow while scanning for beautiful flowers. While walking, he tilts his head to hear the birds and feels a light breeze on his skin. Walking barefoot provides great sensory feedback to the arches of his feet, giving him a good sense of where his feet are in relation to the rest of his body. Listening to the birds chirp helps him orient himself compared to the other creatures in the wild. The light breeze keeps him alert, while the warmth of the sun comforts him. This is the optimal state for sensory integration to occur—when we are aware of our surroundings, but relaxed and calm.

On the other hand, man-made environments, such as movie theaters, colorful play spaces, and indoor party arenas, can overpower the senses and send a child into the fight-or-flight response—an unhealthy state to keep our children in. This time imagine your child is walking into a place where the music is blaring, lights are flashing, and bright colors flood the room. The room is crowded and he keeps bumping into other people. He starts to sweat and gets slightly overwhelmed. Maybe he reacts by covering his ears, talking a little more loudly, or even asks to leave the place, insisting that you never bring him back here again. It is hard to achieve good sensory integration when surrounded by noise and chaos.

On the other hand, nature stimuli tend to be more gentle, subtle, preventative, and, in some cases, even restorative. In fact, nature offers the *ultimate* sensory experience, and we are born as sensory beings. We learn about ourselves and the world around us through our senses, and the more we refine our senses, the better we are at doing...just about everything. Spending some time every day outdoors—from simply walking barefoot on the grass to listening to birds in the trees—offers many sensory benefits:

- **The natural integration of our senses.** Good sensory integration means optimal brain and body performance.

- **A calm but alert state.** When you are in a calm and alert state, you are better able to process the sensory information around you and start to organize the senses, bringing together all the puzzle pieces to form an accurate picture of the world around you.

- **A "just right" amount and kind of sensory stimuli.** Nature doesn't bombard children with too much sensory information at once, which creates a sense of chaos and confusion.

The Outdoors Inspires the Mind

Remember the little girl with the magic wand we talked about earlier in this chapter? Well, let's say she decides to go ahead and build that fort she was afraid to make indoors. Only now she is outdoors. She wants to create a hiding spot away from the fire-breathing dragon. Instead of using blankets and pillows, she uses the surrounding natural environment as her inspiration. She starts by grabbing large sticks and leaning them against a partially fallen tree toward the back of her yard. She notices there are a ton of ferns in the marshy area that borders the woods and starts gathering these, carefully layering them on top of the sticks. She quickly grabs some pinecones and acorns for "food" and crawls into her fort to rest and "eat" away from danger.

The outdoors offers limitless potential to young children. It becomes a place where they can go to relax their mind, to be inspired, and to dive deep into a world of imagination. It's a place where they can design, create, and explore. The possibilities are endless. Time and time again, studies show that when children have free play outdoors, they become better problem solvers and their creativity is enhanced (Hamilton 2014).

Sergio Pellis, a researcher at the University of Lethbridge in Alberta, Canada, says the "experience of play changes the connections of the neurons at the front end of your brain. And without play experiences, those neurons aren't changed" (Hamilton 2014). It's specifically through so-called free play—the kind that requires no coaches, no umpires, and no rulebooks—that lasting changes are made in the frontal brain, which plays a critical role in regulating emotions, making plans, and solving problems. Whether free play is wrestling with your friend or creating an elaborate sand castle together, children have to negotiate and determine their own rules. (See chapter three for a deeper discussion of the benefits of free play.)

Within an indoor environment there are certain expectations and preconceived notions or ideas already created for children. Indoor toys have a designated purpose and, therefore, impose limitations on play. For instance, a puzzle is designed to fit together in a certain fashion. A board game comes with predetermined rules on how to play. A toy car stays a toy car, although the terrain the child creates may vary. However, a pinecone can become something else entirely. I've seen it become treasure, a key, currency, a building material, a decoration, and more. I've seen sticks used as fishing poles and play weapons and as parts of a fort, a boat, a horse, an obstacle course, a trap, and even an airplane.

Daily exposure to the outdoors stimulates the brain in many ways:

- **There are no expectations.** Children are forced to use their imagination in order for that stick, rock, or pinecone to become a part of their world.

- **There are endless possibilities.** The outdoors challenges the mind to constantly think in new ways.

- **There is no pressure.** When engaging in active free play, children can play with others or not, make up their own rules or follow someone else's, be rough-and-tumble or quiet and contemplative.

The Outdoors Offers Risk and Challenge

Imagine a child you know (maybe even your own) in gymnastics class on a balance beam. She walks across it with bare feet. It is consistently smooth, warm, and unbelievably level. She knows its length and feel. There are no surprises. Now imagine her outside, walking barefoot across a log that crosses over a shallow marsh. She walks across the anything-but-level balance beam for a couple of feet just fine. The soft moss tickles her feet and the log is relatively dry and warm. If all of a sudden it gets soft on her, she has to adjust her balance quickly to keep from falling in the marsh that surrounds her. The water from the marsh trickles over her feet and mud gets in between her toes. She experiences a moment of fear and then joy as she realizes she is not going to fall in. She keeps walking, and once off the log she feels the crunchy dried-up leaves of the forest crackle under her feet.

Walking on the log not only ignited all of her senses at the same time, but it also challenged her to react, tested her balance, and required that she learn to persevere in the face of difficulty.

The outdoors is unpredictable, and oftentimes children will come across things that are unexpected. The outdoors forces them to assess their environment and evaluate risks. When a young child becomes adept at evaluating her environment, assessing risks, and accepting challenges, she also becomes confident. As children learn to navigate uneven terrain without falling, figure out how to cross a stream without getting wet, and successfully hike up mountains with their parents, they learn to gather their strength and persist, even when something seems difficult or impossible. They learn what they are physically and mentally capable of when they try and try again.

Evaluating risks and taking challenges while playing outdoors every day is rewarding in many ways:

- **Children build confidence.** When children overcome obstacles, they learn that they can be successful if they keep trying, even in the face of difficulty.

- **Children challenge themselves at their own pace.** When playing outdoors, children get to determine when they are ready to take risks and even control how much risk they are willing to take.

- **They learn to be adaptable.** When playing outdoors, children quickly learn that they can't always control the outcomes of their play. For instance, their fort may not have turned out exactly as they envisioned, and in turn they learn to be flexible in their thinking.

HOW IS NATURE THERAPEUTIC?

By design nature is inherently therapeutic. Everything from the scents of flowers to the sounds of birds stimulate the senses and set children up for healthy sensory integration. We are going to spend some time now learning about how some of the senses are not only enhanced by but literally thrive when children play or even look at natural settings.

Nature is Calming

Adam Alter, an assistant professor of marketing and psychology at New York University's Stern School of Business, describes the phenomenon of how nature is calming perfectly. "Nature restores mental functioning in the same way that food and water restore bodies. The business of everyday life—dodging traffic, making decisions and judgment calls, interacting with strangers—is depleting, and what man-made environments take away from us, nature gives back" (2013).

Letting children play outdoors, away from the hustle and bustle of everyday life, provides respite. It gives them a break from the constant routine, the *Hurry up, we are going to be late* requests; bright colors; noxious smells; and noise and commotion that the man-made world has created. It allows them to unwind and recharge. I regularly observe nature having a calming effect on children. In fact, at TimberNook we've noticed that children are louder and more active when close to buildings versus when playing in the river or woods. Away from buildings, time and time again, children disperse, get quiet, and find purpose.

Even simply looking at nature is calming for children. Researchers asked a hundred sets of parents—whose children presented with attention deficit/hyperactivity disorder—how their children responded to different playtime activities. The children who sat indoors in a room with natural views were calmer than children who played outside in man-made environments devoid of grass and trees (Taylor, Kuo, and Sullivan 2001). This study shows that whether children are indoors or outdoors, having nature present is a key ingredient to grounding and relaxation.

We know now that nature stimuli are calming to children. Although complete immersion in nature, away from buildings, provides total restoration for children and should be done when possible, this may not be practical or even possible for everyone. Simply looking at or being around some nature stimuli will certainly help children to relax. Consider growing a garden for your child to interact with, planting trees, even just having a small area of grass for your child to play on; all these examples provide benefits that blacktop alone can't provide. In addition to providing this access to nature at home, try taking occasional trips to state and national parks to enjoy nature as a family.

Nature Improves the Visual Sense

Nature stimuli are often subtle and mild. The colors found in nature are typically gentle on the eyes. They do not overpower or overstimulate. We learned from the study mentioned above that simply *looking* at nature calms children. As human beings, we rely heavily on our visual sense. What we allow our children to see on a daily basis will affect their mood, temperament, and ability to focus. Also, playing outdoors can positively affect the function and growth of the eyes. The following section will touch upon both of these aspects of children's vision.

SIMPLY LOOKING AT NATURE IMPACTS CHILDREN

My old office was awash in bright colors designed to excite children. However, since most everything in the clinic was screaming *Look at me!*, children became visually overwhelmed in that space.

My oldest daughter, for example, used to love to visit the therapy clinic. However, as soon as she entered the room she lost the ability to effectively regulate her senses. Her visual sense became so overwhelmed that everything else went out the window. For instance, the volume of her voice became increasingly louder, and she ran from one piece of equipment to the next in a hyperactive state. She was literally "off the wall." One reason for this behavior is that our visual sense is designed to alert us to danger, and everything in the clinic caused a heightened state of alertness. Visual overstimulation sent my daughter's arousal level through the roof.

On the other hand, when my daughter played in the woods she had no difficulty controlling her activity level. She was grounded and calm. She was still active, but she maintained control of her body. Why did the natural environment affect my daughter's ability to regulate her behavior? There are a number of studies that have looked at how visual environments impact learning and mood.

Carnegie Mellon's Anna Fisher, Karrie Godwin, and Howard Seltman (2014) looked at whether classroom displays affect children's ability to maintain focus during instruction and when studying lesson content. They found that children in highly decorated classrooms were more distracted, spent more time off task, and demonstrated smaller learning gains compared to when they were in classrooms with blank walls. "We have shown that a classroom's visual environment can affect how much children learn," says Fisher. In other words, keeping things visually simple (as nature has already done for us) can assist with learning.

Simply looking at nature can have an uplifting effect on people. In the early 1980s, a researcher analyzed data collected on patients who had undergone gallbladder surgery between 1972 and 1981. The researcher looked at the recovery rates for patients who had different views from their hospital rooms: some had a view of a brick wall, others faced a small stand of deciduous trees. Apart from their differing views, the rooms and treatment methods were identical. On average, nurses recorded four negative notes for each patient with a view of a brick wall. Comments such as "needs much encouragement" and "upset and crying" were common. On the other hand, patients with a view of the trees warranted negative notes only once during their stay.

Those who had a view of nature also recovered faster, and on average they went home a day earlier than their counterparts. The results of the study are quite substantial because they show that the patients who gazed out at a natural scene were *four* times better off than those who faced a wall (Ulrich 1984).

Man-made environments often utilize colors that are not found in nature. These stronger and more intense visual stimuli can have an alarming effect on our brainstem, particularly the reticular system. The *reticular system* is responsible for processing and integrating sensory information. It contributes to our level of arousal, or alertness. If the visual stimuli are too forceful, children can experience heightened levels of arousal and activity. On the other hand, softer colors and more subtle visual stimuli have a calming effect on a child's sensory system. This leads to an organized and calm state that is ideal for promoting healthy sensory integration (Roley, Blanche, and Schaaf 2001). Children need time being in and looking at nature in order to be in an optimal state for play and learning. Allowing nature to be a visual part of children's lives not only improves their mood and prepares them for learning but sets them up for healthy sensory integration.

PLAY OUTDOORS IMPROVES EYE FUNCTION

Spending time in nature also improves the function of the eyes. As we discussed in chapter one, myopia, or nearsightedness, is at an all-time high. Once thought to be caused by looking at electronic screens for too many hours a day, new studies suggest that children are actually more prone to myopia if they don't spend enough time outdoors. Schools in Asia have found this research so compelling that they are now trying to increase the amount of outdoor time children have daily in hopes of reducing the need for glasses.

Optometrist Donald Mutti (OD, PhD), of the Ohio State University College of Optometry, says data suggests that children who are "genetically predisposed to myopia are three times less likely to need glasses if they spend more than fourteen hours a week outdoors." He goes on to say, "But we don't really know what makes outdoor time so special. If we knew, we could change how we approach myopia" (Ohio State University College of Optometry 2014).

However, scientists do have theories about why spending time outdoors helps reduce instances of myopia. One theory suggests that between the ages of five and nine a child's eyes are still growing. Sometimes this growth causes the distance between the lens and the retina to lengthen, leading to nearsightedness. Scientists feel that outdoor light may actually help preserve the shape and length of the eye during this growth period. They also believe that the bright light of the sun may be a contributing factor—that the pupil of the eye will respond better (open and close more effectively) if exposed to natural light on a regular basis (Ohio State University College of Optometry 2014). In essence, children's eyes will function better when exposed to bouts of sunlight.

Children need to spend time outdoors on a daily basis, not only to be exposed to natural visual stimuli that foster the visual sense and help regulate their moods, but also to support healthy eye function and growth.

Nature Fosters Listening

Loud sirens. Traffic noises. An alarm. Noisy concerts. Blaring music. These types of sounds, also known as noise pollution, often put children into a fight-or-flight response. In such a state they are no longer able to pay attention to what is in front of them (Frick and Young 2012). Our bodies weren't meant to be in a constant state of arousal or stress. In fact, being exposed to noise pollution for hours every day may actually harm young children.

A group of neuroscientists recently performed a study on rats to measure their response to moderately loud to intense levels of noise for more than an hour. They found that prolonged exposure to loud noise actually alters how the brain processes speech, potentially increasing the difficulty one has distinguishing speech sounds (Reed et al. 2014). If noise can alter the brain of a rat, there is a possibility that it can also alter the brain of a human. If this in fact is true, children who are exposed to loud noises on a regular basis may eventually have trouble processing what they are hearing.

In contrast, researchers have found that nature sounds provide a restorative effect. In a study at Stockholm University, forty adults were exposed to sounds from nature and noisy environments after completing stressful mental arithmetic tasks. The researchers found that the sympathetic nervous system recovered faster when the subjects listened to nature sounds versus the noisy stimuli (Alvarsson, Wiens, and Nilsson 2010). If loud man-made noises can alter the brain so it doesn't work as effectively, and nature sounds (for example, sounds of waves crashing or crickets chirping) provide healing, it makes sense to advocate for children to be in natural settings in order to enhance and promote positive sensory integration.

There are also great sensory benefits from listening to birds. Many children in occupational therapy wear special headphones a couple of times a day for a few months if they have trouble with hearing and listening. They listen to specially modulated music, some of which contains nature sounds designed to improve children's mood, attention, auditory processing skills (such as the ability to respond to their name more quickly), social interactions, and activity level. The nature sounds activate the auditory center of the brain, helping children orient themselves to their place in space (Frick and Young 2012).

The results of these occupational therapy programs are incredible. After participating in a listening program, most children display substantial improvements in at least two areas of their life, such as having better sleeping habits, experiencing improved emotional control, and being able to respond to their name quickly and effectively (Frick and Young 2012). One study asked children to draw self-portraits. Before exposure to the prerecorded nature sounds, the children drew pictures with missing noses or arms too low on the body, or their body appeared to be floating in space; the drawings also lacked detail, color, and expression. After three months of participating in a listening program, the same children drew themselves standing on the ground, such as a grassy hill or the beach, with brightly colored details all around them, including plants, dirt, and shrubbery. All body parts were accounted for, many of the faces were smiling, and there was much more detail and color (Frick and Young 2012).

I interviewed Mary Kawar (MS, OT/L), a pediatric occupational therapist who studies the relationships between the vestibular (balance), auditory, and visual systems. She works closely with the developers of a well-known American-based listening program. I asked her if children spent more time outdoors simply listening to birds, would it affect their spatial awareness? "Absolutely!" she answered.

Bird sounds help us orient ourselves to our place in space. For instance, you may hear a bird tweet to your far right, and then another off to the left. These tweets help you locate your position in relation to the sounds coming in. However, since noise pollution really dampens the therapeutic effects of nature sounds, it's best to be away from city sounds in order to reap the most benefits from birds singing. The sounds of nature will work to improve children's sensory development over time.

Nature Enhances the Sense of Touch

I have to admit, watching children play and explore in the large mud puddles at TimberNook has to be one of my favorite things to do. I think this is one of the most meaningful sensory experiences a child can have.

Picture children knee-deep in muddy water searching intently for slimy green frogs. Some children stand on the edge of the puddle, not so sure about getting their feet dirty at first. A little girl grabs a frog and squeals with delight. "I got one! Oh, he is slippery!" Other children gather around the girl to take a closer look.

Meanwhile, the children who are standing on the side and watching the frog scene slowly but surely take off their rubber shoes and immerse themselves in the puddle. "Yuck! This feels mushy," a girl says as she processes the new sensations and learns to navigate the water without falling. Another child slips and falls into the puddle. "Ugh!" He is momentarily shocked. No one reacts to his fall. He gets back up and starts laughing. "Look at me! Look how dirty I am!" Another child laughs and purposefully falls into the puddle.

The sensations of getting dirty and messy in real mud offer children an invaluable rich and tactile experience. The tactile system is flexible, and through exposure to various tactile experiences, children increase their tolerance to different touch sensations. If a child has a poor tolerance to touch, as we learned in chapter two, he may have trouble wearing a variety of different clothing, may refuse to go barefoot, and may even have trouble with school tasks such as using glue without getting upset. That's why it is important to expose children to a variety of different textures at an early age.

It is one of the many reasons why there's a growing trend for parenting blogs to include sensory activities—from shaving-cream play to creating slime and differently textured playdough. These experiences are fun and entertaining to children for a short period of time; however, playing outdoors often expands the touch experience to involve the whole body, further enhancing the sensory benefits.

For a better understanding of this notion, let's compare playing indoors with a bin full of sand to being on the beach. The child indoors will most likely engage only his hands in the sensory bin. He may play with plastic scoops and containers. He'll probably sit under the watchful eye of an adult.

Now picture this same child on the beach. The sensations of the sun warm the boy's skin. Water splashes on him and his feet sink in the mud as he fills the bucket with cold, crisp water. He kneels down next to his sand castle—exposing more of his body to the rough sensations of sand. He digs with his fingers to make a moat around the castle. He finds slimy seaweed and rough, spiky shells to line his castle with. By the time he is done building his sand castle, *hours* later since he took breaks to swim and eat, he is covered from head to toe in sand and mud and has a huge grin on his face.

Although the first experience may be what we think of when we say "give the child a sensory experience," in the second example *many* more senses were ignited. The sensations of temperature and exposure to different tactile experiences (slimy seaweed and rough shells) only expanded his sensory repertoire. His whole body was exposed to the sand, not just his hands.

Also, when children are pushing, pulling, or digging on the beach, they are better tolerating and integrating light touch experiences, such as a soft wind blowing on the face and the feel of the sand. Children with sensory-processing issues can sometimes be extremely averse to light-touch stimuli, such as playing in the sand or having seaweed brush up against the skin, when they are experienced in isolation. However, the bigger movements of playing on the beach help override the light-touch sensations and improve tolerance (Ayres 2000). Beach play versus playing with a bin of sand may also be more meaningful to young children, motivating them to play for a longer period of time and to get creative with castle design.

Not only does getting dirty and muddy outdoors increase tolerance of touch experiences, but it also improves the immune system.

HYGIENE HYPOTHESIS

Mud. Wet and sticky mud. Most children can't resist it. Getting dirty and even sampling a little bit of dirt doesn't hurt. In fact, it can be downright healthy for your child. Exposure to dirt, animals, and germs from an early age on can actually improve the immune system. We've long known that children who grow up on farms tend to not have asthma, tend to have fewer allergies, and are less likely to have autoimmune disorders than children who grow up in urban areas (Brody 2009).

Researchers are finding that overuse of sanitizer, cleaning every surface in the house, taking a bath every day, sterilizing all baby equipment, and washing hands always before eating is hurting our immune systems. They call this the *hygiene hypothesis.* The US Food and Drug Administration reports that "the immune response is derailed by the extremely clean household environments often found in the developed world. In other words, the young child's environment can be 'too clean' to pose an effective challenge to a maturing immune system" (2015).

According to the hygiene hypothesis, the problem with extremely clean environments is that they fail to provide the necessary exposure to germs that strengthens the immune system so it can protect us from infectious organisms. Instead, its defense responses become so inadequate that they actually contribute to the development of asthma and allergies (Okada et al. 2010). Therefore, in order to develop a strong and healthy immune system, it is essential for children to be exposed to the outdoors—especially if that exposure involves getting dirty.

GOING BAREFOOT

Not only does going barefoot in nature help to develop and fine-tune touch senses in the feet, it also strengthens the arches. Consider this evidence. Doctors at a medical college in India, where most children in rural towns typically don't wear shoes, noticed that children from the more rural towns rarely presented with flat feet. Most of their clients who had flat feet were from urban areas. They decided to look into this further by analyzing the static footprints of 2,300 children. They found that children who wore shoes were significantly more likely to have flat feet compared to those who did not. They also found that the critical period for the development of the arch occurred before the age of six. Their study suggested that wearing shoes in early childhood is detrimental to the development of a normal, or high, medial longitudinal arch (Rao and Joseph 1992).

When researching the importance of going barefoot, I came across Katy Bowman, a biomechanical expert and founder of the Restorative Exercise Institute. She has this to say about young children wearing shoes most of the time: "Shoes alter human movement. Many of the ailments we suffer from, musculoskeletally speaking, are a result of our dependence on footwear and the strain on the ligaments and plantar fascia from decades of muscle atrophy. If you can start a kid off with a preference to minimal footwear, it saves time and degeneration" (Crawford 2013).

When my youngest daughter was really young, she presented with flat feet. She even needed physical therapy and wore braces for a while to support her ankles and arches. However, it wasn't until she started going almost completely barefoot in the summers that we began to notice a real change. She now has nicely developed ankles and arches and walks fluidly.

Walking outdoors offers natural messages to children's feet as they walk on different-sized pebbles and uneven ground. The resistance and inconsistency nature offers integrates reflexes in the foot and forms strong arches. Going barefoot out in nature helps to develop normal gait patterns, balance, and tolerance of touch in the feet, all of which provide a strong foundation for confident and fluid movement.

Nature Enhances the Sense of Taste and Smell

Taste. Just the word makes the mouth water. Taste is a sense that tells us something about our environment. Young babies put things in their mouth to gather information about the environment. I've seen plenty of moms prevent their children from putting pieces of nature, such as dirt, leaves, and pinecones, in their mouth. Instead, babies today are offered plastic rattles and toys that are engineered to offer sensory feedback. However, nature provides a variety of taste and texture experiences that are hard to replicate in toys or other man-made items.

Do you have a child who's a picky eater? Many children today have a decreased tolerance for exploring new textures in the mouth, which often correlates with having trouble trying new foods. Other children have trouble with awareness in and around their mouths. Letting children put dirt or pinecones in their mouth at an early age won't hurt them. In fact, babies learn about their environment through their mouth first. Doing so increases their tolerance of new sensory stimuli in and around the mouth. (However, be careful that they don't put small objects, like rocks and acorns or even animal droppings, in their mouth.) The safety implications of this will be discussed in detail in chapter eight.

Not only does exploring natural items with the mouth improve tolerance and oral sensory awareness in children, it's also good for improving their immune system (as we started to learn about with the hygiene hypothesis discussion earlier in this chapter). In her book *Why Dirt Is Good*, microbiology and immunology instructor Mary Ruebush suggests that when a child puts an object in his mouth, he is allowing his immune response to explore his environment. Not only does this allow for the "practice" of immune responses, which is necessary for protection, but putting objects in the mouth also plays a critical role in teaching the immune system what stimuli are harmful and best to avoid (Brody 2009).

For older children, gathering food (berries, nuts, fruit) right out of nature, as our ancestors did, offers a richer sensory experience than buying these foods after they've been sitting in a supermarket for weeks. For instance, biting into a fresh apple is not only healthier but also a juicier, more flavorful, and even noisier experience than eating an apple from the grocery store. These sensations enhance the sensory experience, therefore enhancing the sensory memories.

Not only are children today presenting with decreased tolerance to touch, many kids are not tolerating different scents. Some kids even gag when they smell new things. Nature offers many benefits to our sense of smell. When you're in nature, you don't smell just one scent but rather a multitude of scents with varying intensities. These tell us information about the environment around us. Indoors, the scents are more constant. Indoors you'll find more man-made and potentially harmful scents, such as cleaning chemicals or fresh paint. Outdoors, you may smell the scents of fall, farm animals, or freshly cut grass, none of which are harmful to our senses.

In fact, natural scents are often used in therapy. This practice is called *aromatherapy*. The essential oils derived from plants and other things found in nature stimulate small receptors in the nose, which then send messages through the nervous system to the limbic system—the part of the brain that controls emotions. Natural aromatherapy smells can relax patients and provide them with a sense of tranquility (University of Maryland Medical Center 2011). In other words, simply letting your children smell different things in nature will help them regulate emotions.

Outdoor Experiences that Engage the Senses

The following examples are specific outdoor experiences for children that ignite many of the senses.

PROMOTE BAREFOOT BABES

Let your children go barefoot as much as possible both indoors and outdoors. If they have to wear shoes, consider slippers or minimalist shoes that allow the arches of the feet to receive input from both natural and man-made surfaces.

GO FRUIT OR BERRY PICKING

Many farms offer seasonal pick-your-own edibles, from blueberries and strawberries to pumpkins and apples. Not near a farm? Try food-oriented festivals and farmers markets, where children can sample different items while having fun outdoors. Making a pie or muffins with your children using the fresh berries they picked is a meaningful sensory experience. Children not only learn where food comes from but they engage their senses of smell, sight, hearing, taste, and proprioception as they mix, measure, and taste the batter.

GARDEN WITH CHILDREN

Gardening with children offers many sensory benefits. They get to dig in the dirt, use a watering can, nurture living plants, taste fresh foods, learn to tolerate new textures, and broaden their food horizon. Smelling the herbs and flowers they grow is also a great olfactory (smell) experience.

GO BIRDING

Identifying different bird sounds is a great auditory (sound) skill. Offer children an Audubon bird identification book so they can learn to look up the birds that they hear and see. Teach them how to make birdcalls using only their hands and voice.

PLAY IN THE DARK

Playing a game such as hide and seek in the dark offers a challenging sensory experience. Many children rely heavily on their visual system to navigate their surroundings. When this is taken away, the balance and proprioceptive (joint and muscle sense) systems need to work harder to keep children upright and coordinated as they travel through the dark. Also, lying down and hiding during a game of hide and seek gets children up close and personal with leaves, dirt, and other tactile experiences. Furthermore, because they are in play, they will often tolerate things they wouldn't normally tolerate (lying down on wet leaves, playing in the dark), especially when playing with other children.

INTERACT WITH ANIMALS

Caring for animals both big and small exposes children to a lot of different textures, smells, sounds, and sights. Therapists have used dogs and horses and other animals for many years in order to work on all sorts of physical, emotional, and intellectual skills with children. Farm animals, such as alpacas, sheep, cows, goats, chickens, and pigs, offer children a variety of sensory input. However, even a cat or a hamster can offer great sensory experiences.

PLAY AT THE BEACH

The beach is a whole-body experience that stimulates the senses of touch (sand, water, varying temperatures), proprioception (digging in the dirt), hearing (birds, crashing waves), seeing (scurrying crabs, landing seagulls), and vestibular (reaching down to fill a pail with water, running on the soft sand).

ENCOURAGE TREE CLIMBING

Tree climbing is a great way for children to learn how to assess risk and their own abilities. As a parent, this may scare you. But fear not, it is important for children to learn these skills before taking bigger risks, such as getting behind the wheel of a car. How does climbing challenge children? As children climb, they learn to check the branches to make sure they are not dead, broken, or unstable. They climb only as high as they feel comfortable.

Oftentimes, children will only climb a few feet off the ground before wanting to come back down. They aren't yet ready for the challenge of going higher. However, through practice and more practice they learn what their bodies are capable of and how to assess their environment for potential hazards—both important life lessons for children.

COOK OVER AN OPEN FIRE

Cooking over an open fire is often a very meaningful and delightful experience for children. I have found that many children are willing to try new food when it is cooked over a fire. The sheer thrill of participating in the preparation of food inspires many children to at least try the food they worked so hard to create. They learn to have patience when cooking over a fire and enjoy new smells as well.

IMMERSE YOUR CHILD IN NATURE

Plan a family vacation to a state or national park where every member of the family can recalibrate and rejuvenate in nature for at least three days. Just as a daily dose of nature is important, so is the occasional total immersion.

IN A NUTSHELL

While man-made environments may excite children, they may overwhelm or overstimulate them. Indoor environments can also understimulate and offer few sensory benefits to children. The great outdoors, on the other hand, offers limitless possibilities for play experiences and exploration of the senses, enhancing and refining the senses through repeated practice. It is through daily play outdoors that your children will challenge and strengthen their senses of touch, vision, hearing, smell, taste, and much more!

CHAPTER 5

"Safety First" Equals Child Development Later

When my daughter was a young toddler, I constantly chased after her trying to make sure she was "safe." I baby-proofed every inch of my house—and grandma's. I was her shadow when she walked, just in case she stumbled. With my ever-present pack of wipes, I sanitized every eating surface we came in contact with. She wore sun hats and I lathered her with sunscreen. I bought the highest-rated car seat.

While I felt like I had a handle on keeping her safe, I realized I wasn't prepared for how active she was! *Oh man,* I thought one exhausted morning, *playdough only lasted five minutes! What do I do now?* That's when I decided we needed to have a plan *every* day. This girl needed to be entertained!

Soon came two-hour playdates, organized sports, and preschool at the wee age of three years old. We got busy, *really* busy—driving to preschool, gymnastics, soccer, music classes, and a moms' group. I don't recall taking my daughter outside to play that much. If I did, it was to the neighborhood playground or a rare trip to the beach. Never did it occur to me to explore the twelve beautiful wooded acres of our backyard with my daughter. In my mind, we were too busy to sacrifice our precious time in nature. It wasn't until later that I realized I had made a big mistake. When she presented with bouts of anxiety, aggressive tendencies, and sensory issues, I knew we were running away from the one thing that would actually help my daughter: time alone in nature.

There is something about letting children explore the outdoors on their own that intimidates many adults. Fear is often the biggest barrier to giving children some wiggle room away from constant adult supervision. Fear comes in many forms. Adults are afraid of childhood abductions, kids getting lost, and kids getting hurt. There are also smaller fears, such as bug bites, wild animals, and poisonous plants, that make parents generally hesitant about play outdoors.

Many of these fears are a result of society exaggerating dangers and parents' loss of trust. In this new era of parenting, we are doing everything we can to protect our children. However, sometimes too much protection can cause more harm than good. We are keeping children from attaining the very life skills and sensory awareness they need in order to grow into resilient and able-bodied adults.

In this chapter I will address the most common fears related to letting kids play independently outdoors and take risks. I will also address why independent play and risk taking are actually critical to healthy child development and discuss safety tips for when you let your child head out to play.

WHAT ADULT-DRIVEN SAFETY LOOKS LIKE

There is a huge difference between what most parents fear (abduction by strangers, serious life-changing injuries) and what's actually happening. Tragedies such as a child being kidnapped or dying from a playground accident are incredibly rare. They were rare thirty to forty years ago and are still just as rare. What has changed is our trust in human nature, causing us to protect what we can—especially our children.

This tightening of the reins has resulted in an era of supervised playdates, an increase in organized activities, and a reduction in the number of children playing outdoors. Does that sound like the routine in your home? At first glance, this may not seem harmful. However, the lack of time to explore and play independently outdoors is affecting social skills, emotional security, independence, and creativity. It is even creating children who are more accident-prone and susceptible to harm.

Constant Supervision

"In all my years as a parent, I've mostly met children who take it for granted that they are always being watched," writes Hanna Rosin, author of the wildly popular article "The Overprotected Kid" (2014). This sentiment reminds me of the classic book *1984*, in which Big Brother is always watching—there is no escape. The characters in the book are under constant surveillance. The same can be said of children today. They are under relentless supervision. If you see a child in public, usually there is a parent only a few feet away. I rejoice when I see a child playing outdoors without a parent in sight, for in this day in age, such a sight is a rarity.

Fear of Strangers

One of the biggest culprits for the incessant management of children's whereabouts is our fear of strangers. If you ask most parents why they don't let their kids out to play alone, many will answer, "It is a different world out there. There are too many creeps." The truth is, the world really isn't any more unsafe now than it was in the 80s and even the 70s. If anything crime has decreased. David Finkelhor, director of the Crimes Against

Children Research Center and one of the most reliable authorities on child abduction, has found that crimes against children have actually decreased since the 1990s. The type of abduction in which a stranger takes a child remains extremely rare and has not increased in frequency (Skenazy 2009).

The only type of abduction that is increasing in number is family abduction. These abductions, usually the result of a custody battle, are when either the mother or father kidnaps the child. Sometimes these types of abductions are lumped together with stranger abductions in FBI crime reports, creating alarming statistics (Rosin 2014). Headlines in the news read, "Child Kidnappings and Abductions Could Be Four Times Higher Than Authorities Admit, Charities Warn" (Fearn 2015) or "Child Kidnappings and Abductions Soar by 13 Percent" (Russia Today 2015). These headlines are meant to grab our attention so we'll read the paper or listen to the news. And when a stranger does actually abduct a child on one side of the country, the whole country hears about it—often for weeks. This fuels our fear and makes us think that it is likely to happen to our children too, which is one reason why we keep them under constant supervision and busy with organized indoor activities.

Right to Roam

Due to "stranger danger" fear, we are keeping closer tabs on our children than ever before. This surveillance includes how far we allow children to walk alone or whether or not they play by themselves. My daughters are seven and ten, and we are just now letting them bike down our dirt road to play with children in a connecting neighborhood. All in all, it's about a half-mile trek. This allowance has earned us the title "free-range parents" by many of the families we know. "Free-range kids" is a term coined by Lenore Skenazy—a journalist turned author—in her book of the same name. She is doing her best to advocate for children's rights to play and roam neighborhoods and streets on their own.

My husband recently told me that we are much stricter with our girls than his parents were with him in the 80s. When he was twelve, he and his younger brother (10 years old) and a few other friends biked twenty-five miles to the nearest big town on their own. They spent the day in town roaming the streets, window-shopping, and buying candy. They did this once a year, for four years in a row. On the way they'd stop and play with friends who lived ten miles away before heading back out on their adventure. He smiles at the memory and takes pride in the fact that he did this at such a young age. "If we let our girls do this at ten and twelve years old, we'd probably be put in jail," he says, clearly frustrated.

It is true. It is coming to a point where we may no longer have the ability to let kids have as much freedom as parents did in the past. Even though the world is no more dangerous than when we were growing up, some people now overreact upon seeing children outdoors on their own. Two Maryland parents were under investigation twice for child neglect just for letting their ten-year-old and six-year-old walk together to the park a mile away. The children were picked up by the police and the parents were questioned. What child protective services felt was neglect, the highly educated parents (one is a fiction writer and the other a physicist) felt was "absolutely critical for their development—to learn responsibility, to experience the world, and to gain confidence and competency" (St. George 2015). Unfortunately, allowing children the freedom to roam is seen by some as poor parenting.

A recent study in the United Kingdom found that children's "independent mobility" is decreasing. In 1971, 86 percent of English primary schoolchildren were allowed to travel home from school alone. By 1990, this number had dropped markedly to 35 percent. By 2010, only 25 percent of children walked home without an adult. When compared to their German peers, English primary schoolchildren had less independent mobility in 1990, and this remained the case in 2010. Children in Germany were granted all the licenses of independent mobility in greater proportions and at earlier ages than their English counterparts (Shaw et al. 2013). This demonstrates that depending on societal norms, children's ability to be independent and take risks varies widely. Tim Gill, childhood researcher and author of *Rethinking Childhood*, states, "I've been told that in Switzerland, parents are judged badly if they DON'T let their children walk to kindergarten (yes, kindergarten) on their own" (Greenfield 2015).

By keeping children under constant supervision, we are keeping them from attaining the very freedom they need in order to move their bodies fully and frequently throughout the day. The more children play outdoors, the more opportunities and space they have to challenge their bodies. From having to bike back and forth between friends' houses to playing a game of pickup baseball for hours to enjoying flashlight tag at night, outdoor activities get children moving and playing.

When given the opportunity to play on their own, children demonstrate greater levels of confidence and improved social skills. There are also wider community benefits of independent play, such as closer neighborhood relations, a stronger sense of community, and less fear of crime (Shaw et al. 2013). Children also get to have rest away from the adult world, a sense of privacy, and an opportunity to keep secrets. They get to create their own worlds and top-secret hideaways and forts, take shortcuts that adults might not necessarily take, and even invent their own languages. All of these benefits of independent play foster a sense of responsibility, camaraderie among peers, independence, and imagination, and they enhance children's play skills. Children simply need the space to roam and the chance to play on their own.

Fear of Injuries

"Be careful," says a mother walking closely behind her three-year-old on a bumpy nature path at TimberNook. "Don't fall," the mother says. Moments later she warns again, "Don't fall." *How many times have I done this to my own children?* I think to myself. They get too close to the edge of a rock and my own protective instincts come out. "Be careful," I say with dire caution in my voice. Then they walk on something wet and slippery and I automatically advise again, "Be careful." But honestly, what's there really to be careful of? What are we so afraid of?

Most of us are worried that our child will fall and get hurt in a serious way. In our minds a fall can become a devastating or even fatal mistake. However, children are often more capable of assessing risk than we give them credit for. And in reality, a fatal accident is extremely rare. And if they do fall? Many times a bruise or a scrape serves as a far better learning experience than a parent repeating "Be careful!" every few minutes.

"The problem is that the public assumes that any risk to any individual is 100 percent risk to them," states Dr. F. Sessions Cole, chief medical officer at St. Louis Children's Hospital (Skenazy 2009, 7). For instance, if a child falls off a jungle gym and gets a serious head injury, we automatically think that our child is at risk for the same thing. We may even have our children avoid jungle gyms altogether, sacrificing the physical benefits of the equipment in the name of "safety."

During the past thirty years, playgrounds have changed drastically for this very reason. In the late 1970s, the parents of a few children who were seriously hurt on playground equipment brought forth lawsuits. The lawsuits led to changes in playground policies and a new era of ultrasafe but unchallenging playground equipment. (Playground changes will be explained in further detail in chapter six.) However, the changes to playground equipment and the addition of softer padding under the equipment have not made a difference in the number of injuries. In fact, the number is actually on the rise.

Parents may think that being extra cautious and protective of their children is worth fewer injuries. The irony is that our close attention to safety has not made a difference in the number of children getting hurt. In fact, according to the National Electronic Injury Surveillance System, which monitors hospital visits and the frequency of emergency room visits due to playground injuries, there has been a steady *increase* in the number of playground injuries since the early 1980s—when most of the equipment renovation started. In 1980, the number of visits resulting from playground accidents was 156,000. By 2013, the number had jumped to 271,475 (Rosin 2014).

Severe head injuries, falls, and deaths at playgrounds remain extremely rare—regardless of how much safety proofing is implemented. The Consumer Product Safety Commission recently reported that there were one hundred deaths due to playground injuries between 2001 and 2008. Out of the millions of children in the United States, this equals about thirteen children a year— only ten fewer children than reported in 1980 (Rosin 2014).

Falling from time to time and experiencing reasonable risks (climbing up rocks, riding a bike to a friend's house, playing in the dark) actually benefit healthy physical development. When children are given the opportunity to take risks and even experience the sensation of falling, they learn how to make necessary motor adaptations over time, such as shifting body weight on a bicycle to keep from falling or reaching out with a hand to protect the face during a fall. Adults can't teach a child how to make these necessary adaptations; children have to learn these skills through real-life experiences (Ayres 2000). They need to fall every now and again, make mistakes, and even get bumps and bruises in order to develop good balance, become more reliant, and develop effective motor skills.

Limiting children's exposure to risk and constantly trying to keep them from falling can impede their physical development. They can actually become unsafe. Without the freedom to make adaptations to the motor and balance system, children may be clumsier and more likely to fall and be seriously hurt. A child may trip and forget to put out his hand to protect his head from injury. A child may look over her shoulder for a second while on top of a rock, which causes her to fall since she hasn't yet developed the motor skills that allow her to do two things at once. Instead of running away from all risks, we need to let children experience gradual risk to develop the essential physical skills they need to stay safe.

An Abundance of Rules

American society has become litigious over the years, and it is really starting to take a toll on child development. Schools and even towns are implementing more rules and banning classic childhood pastimes like playing tag, sledding, playing on monkey bars, and jumping off swings. The reason? Some child somewhere got hurt, and now the school or town is fearful of being sued. The easy solution is to get rid of the equipment and forbid the activity. This may be a good short-term solution; however, we have forgotten the importance of allowing children to participate in these activities.

A friend of mine who is an elementary school teacher tells me that she had to give up being a recess monitor. "I just couldn't take it anymore," she says. "I had to tell the kids no to things that I knew were good for their development." For example, the kids were highly restricted in how they could use the swings. "I remember jumping off swings when I was a kid to see how far I could get.

Now they are not only forbidden from jumping, but they aren't allowed to do anything but sit in the swing. I even had to tell them they couldn't swing on their bellies," she states. I've had children tell me they are not allowed to spin in circles on the swings and other teachers tell me that kids are not allowed to go on top of the monkey bars.

As you learned in chapter three, spinning in circles and even going upside down is important for establishing a strong balance system. Climbing on top of monkey bars is a great physical challenge, and activities such as sledding strengthen the core and give necessary vestibular input for good body awareness. If we don't allow our children the freedom to move their bodies in different ways with these simple challenges, how do we expect them to become capable of navigating their environment without getting hurt? Risks and challenges are important for healthy sensory and motor development.

When I was helping to start a TimberNook in New Zealand, I took time to meet Bruce McLachlan, the principal of Swanson Elementary School in Auckland. Bruce did the unthinkable. As part of a local university study, he got rid of rules during recess time. Kids were allowed to scale fences, climb trees, build with construction materials, and ride scooters. A funny thing happened. There was a decrease in the amount of bullying, a drop in the number of serious injuries, and an improvement in concentration in the classroom.

Bruce told me that part of the reason why getting rid of the rules worked in New Zealand is that he doesn't have to worry about litigation. New Zealanders benefit from a state-run universal insurance plan administered by the Accident Compensation Commission, which covers medical costs if people injure themselves. Because of this, it is rare for parents in New Zealand to sue a school. In fact, one little boy broke his arm during recess on a scooter. Bruce recalled this experience to me. "The father called me asking if he could come and talk with me. I was thinking the worst. But you know what he said? The father came and stated, 'Thank you. Please don't change the rules for recess because my son got hurt. It is an important life lesson.' I was truly shocked."

The story about Bruce eliminating the rules during recess has gone viral. Adults everywhere are now thinking back to their childhood, when they were allowed more freedom. To them, this story represents a fight against the era of bubble-wrapping our children and instead focusing on the value of free play.

WHAT CHILD-DRIVEN PLAY LOOKS LIKE

Children are naturally curious and seek out opportunities to make sense of the world. When children are left to their own devices, they experiment with their surroundings, take risks, make mistakes, and then learn from the mistakes. They problem solve, negotiate, imagine, and investigate. Children learn an immense amount of information through free play. Our children will be well served if we offer them the freedom to play on their own, to learn through making mistakes, and to come to their own conclusions about the world around them. This process will help prepare them for life while improving their cognitive, social-emotional, and physical skills at the same time.

Children Know What They Need

As adults, we may feel that we always know what is best for our children. A child's neurological system begs to differ. Children with healthy neurological systems naturally seek out the sensory input they need on their own. They determine how much, how fast, and how high works for them at any given time. They do this without even thinking about it. If they are spinning in circles, it is because they need to. If they are jumping off a rock over and over, it is because they are craving that sensory input. They are trying to organize their senses through practice and repetition.

I've heard many adults and even so-called experts claim that spinning can be dangerous. "Children shouldn't spin," an adult may claim. "They may feel sick afterward." To an extent this is true. However, it can be risky for an adult to control how much sensory input a child gets. Adults can accidently send a child into sensory overload if they don't know what the warning signs are when a child is not tolerating vestibular input. *Sensory overload* is when the sensory input becomes too much for a child; it can cause the child to feel sick for the rest of the day. Another downside of sensory overload is that the child may refuse to do that activity again in the future. However, the danger goes away when the adults step back and let the child determine the adequate amount of sensory input.

You may find yourself having a natural tendency to say, "No spinning, you may get dizzy," or "Get down from that tree, you might get hurt." However, when we restrict children from experiencing new sensations of their own free will, they may not develop the senses and motor skills necessary to take risks without getting hurt. Then we, the adults, become the barrier to healthy child development. Later, when the children are older and get behind the wheel of a car, they may not have the skills necessary to safely navigate the roads. It's important for us to give children opportunities to practice moving their bodies in new ways at an early age; doing so will prepare them to be safer in the long run.

Children Were Born to Take Risks

Children are natural risk takers. They need it. They crave it. Ellen Sandseter, a professor of early-childhood education at Queen Maud University College in Trondheim, observed and interviewed children on playgrounds in Norway. She found that children have a sensory need to taste danger and excitement. She defines risky play as "thrilling and exciting forms of play that involve a risk of physical injury." She identifies six types of risky play: (1) handling dangerous tools, such as knives and hammers; (2) being near dangerous elements, such as fire and water; (3) exploring heights, by climbing trees and rocks; (4) speed, such as skiing fast down a hill; (5) rough-and-tumble play, such as wrestling and play fighting; and (6) playing on one's own. She feels the last one is the most important for fostering healthy development (Rosin 2014). "The urge to walk off alone in new and undiscovered environments without supervision from adults is children's way of exploring their world and becoming at home in it" (Sandseter and Kennair 2011, 269).

Most of the time risky play occurs during free play, as opposed to play organized by adults. During free play, children learn to manage, control, and even overcome their fears by taking risks. Sandseter refers to risky play as a form of *exposure therapy*, in which children force themselves to do things they're afraid of in order to confront their fears and overcome them. For instance, a child may launch her bike off a jump built of wooden planks or perform new tricks on his skateboard. According to Sandseter, this type of risky play provides a desensitizing effect. Through research she has found that children who injured themselves by falling from heights between ages five and nine are less likely to be afraid of heights when they are eighteen years old (Sandseter and Kennair 2011).

On the other hand, if children never go through the process of exposing themselves to new risks, their fear can turn into a phobia (Rosin 2014). Sandseter states, "Our fear of children being harmed by mostly harmless injuries may result in more fearful children and increased levels of psychopathology" (quoted in Tierney 2011). In other words, our parental anxiety can become a barrier to children's emotional development. JoAnn Deak (PhD), author of *Girls Will Be Girls*, states, "Girls who avoid risks have poorer self-esteem than girls who can and do face challenges" (PBS Parents n. d.). Therefore, in order to help deter the rise in social-emotional issues we are witnessing in children today, we would be wise to provide thrilling play experiences for children.

Children Take Pride in Independent Play Experiences

The third-grade classroom that was visiting TimberNook for the day consisted of mostly boys—rowdy, loud, and rambunctious boys. As soon as the children realized they had the freedom to explore and build in the woods, something funny happened: they got really quiet. They dispersed, and many of them started working together to build a massive teepee.

Nothing gives me more pleasure than to see children contentedly building a structure using branches, bricks, and logs out in the woodland. That is, until fear kicks in and everyone's pulse increases a few notches at the cry of alarm.

"Put the sticks *down!*" I looked over to see a chaperone running frantically toward the children. "Danger! Danger!" she shrieked. Momentarily astonished by the sudden state of perceived emergency, I finally found my voice. "It's okay," I reassured her. "I said they could use the sticks as long as they respect each other's personal space." Speechless, the chaperone frowned, turned, and walked to a group of nearby chaperones. I could have stopped the kids from building, given in to the fear, and encouraged them to do something considered a little less risky by the surrounding adults. However, I decided to let the kids proceed.

The children, with the help of a few excited adults, persisted with building their colossal stick teepee. "Look at what we built!"

one of the boys said proudly, showing off their work. "Can you believe it?" another child eagerly asked.

During this time of construction, no child got hurt, which would have been fine. Getting bumps and bruises is a normal part of healthy outdoor play experiences—a right of passage of sorts. But not only was no harm done while these kids participated in "risky" play, but they took great pride in the work they completed.

Letting children take risks boosts their confidence. Using a knife to whittle a stick, exploring without an adult, tending a fire, and creating a fort all have one thing in common: there is the risk of injury. Even though letting kids take risks can be scary for parents, these experiences offer considerable reward and value to growing children.

When a child takes a risk, such as riding a bike for the first time, it can be frightening. At the same time, the child is learning to overcome that fear to reach a goal. In the process of learning to ride a bike successfully, the child learns patience, perseverance, and resilience. She learns to keep trying, even when she continues to fall. In the end, she will be riding her bike well. She may think, *I did this all on my own.* What a great life lesson, because life is a continuous series of trials. Having the confidence to persevere and have patience helps children to be successful with relationships, school, and work experiences later in life.

Taking Physical Risks Improves Safety Awareness

Not only does taking risks help children overcome their fears and build confidence, but doing so also helps children develop strong physical skills that support good body awareness. Having adequate body awareness is essential for the safe navigation of and interaction with the world around us. There are many simple activities that are excellent for promoting balance and good body awareness:

- Spinning in circles

- Rolling down the hill and back up again

- Dancing

- Gymnastics

- Skating (especially when going backwards, spinning, and turning)

- Playing on the merry-go-round

- Going upside down

- Swinging (on the bottom or belly and while standing or spinning)

- Swimming

- Diving

- Climbing

- Crawling

COMMON SENSE SAFETY IN THE OUTDOORS

No matter where you live, you'll want to teach your child how to safely get around on his own. If you live in the city, teach your child street smarts. He needs to learn how to read the signs, how to get from point A to point B without getting lost, how to cross the street safely, who to talk to in case he needs help (for example, to look for a police officer), how to stick with his group of kids, and so on.

If you live in the suburbs, you may change the lessons slightly. You may focus on teaching your child how to safely ride her bike to the neighbor's house, to check in with you to let you know she is going somewhere new, how to watch out for cars, how to cross the street safely, to not go into the houses of strangers, and so on.

If you live in the country, you may focus on teaching your child what poisonous plants to look out for, what to eat and not eat in the wild, how to safely navigate the woods, where the boundaries for playing are, that he's not to go swimming in the river without an adult, and so on.

Below are a few safety tips to consider when children are playing in more natural settings.

Cuts and scrapes—If a child gets a cut, bruise, or scrape, it is best to not make it a big deal. Children look to us to see if a situation is safe or not. If we model that getting small boo-boos is simply a normal part of playing outdoors, they will come to realize that bruises, cuts, and scrapes aren't really that big of a deal. Keep bandages handy just in case. Clean the cut or scrape and place a bandage on it only if it's bleeding. Otherwise, just clean it and let it heal.

Getting dirty and wet—Your reaction should be similar if a child slips in mud or water and accidently gets dirty or wet. Again, if you don't make it a big deal, your child will learn that there is nothing to be anxious about. Reacting to something like this with alarm does not help the situation at all and is likely to overwhelm the child. Keep a spare change of clothes or towel nearby or in your car for quick changing if necessary.

Poisonous plants—It is important to teach children which plants are poisonous and which ones aren't. I've encountered many kids who thought every plant they encountered was poison ivy, and this lack of knowledge created great fear. They were afraid to even walk off a path in the woods. It is best to show children what is and isn't poisonous so they aren't fearful of every plant.

Edibles in the wild—If your child is older and can tell the difference between plants, it may be fun to show her the plants that are edible in nature. My girls often pick wintergreen in the wild, and then we make wintergreen tea together. They love finding out that there is stuff right outside our door that they can munch on. It is also important to show your child the plants she should never put in her mouth, and to teach her that if she is not sure if something is edible or not, to just leave it.

Staying hydrated—It is so easy for little ones to get caught up in the fun of playing outdoors that they forget to drink. Make sure your child constantly has access to water. A good way to motivate children to drink is to add a little flavor to the water, such as frozen fruit or just a bit of 100-percent juice.

Bugs—Educate yourself first about the bugs that you need to watch out for before discussing them with your child. If it is really "buggy" outside, consider putting bug spray on your child. There are all sorts of natural bug sprays. Learn about the bugs to be wary of, but don't let them keep you and your family from exploring the great outdoors. Stay calm if your child gets a bite from something that concerns you. The last thing you want to do is make your child afraid to go outside again. For instance, when my children have ticks on them, I calmly remove them and clean the area. We check for ticks every night, so the girls know that we will always make sure they are "tick free" before they go to bed.

Sun exposure—Playing outdoors is important for vitamin D. However, if children are outdoors for extended periods of time where there is little shade, consider putting sunscreen on them to protect from burns. You can also purchase special clothing that protects against the sun. Wearing hats also helps prevent burns to the head.

Getting lost—Whether you live in the city, the suburbs, or the country, teach your children how to get around their outdoor space well. This may entail accurately reading maps, a compass, a trail, or signs. Having physical boundaries usually helps prevent children from getting lost. For instance, you may say, "You can go as far as Tony's house before you need to turn around." Also, it is important to teach them what to do in case they do get lost—for example, who to ask for help.

Wild animals—Research the animals in your area and any necessary precautions you should take. Teach these precautions to your child without scaring him by continuing to bring up facts he already knows. Simply tell him what to do if he crosses paths with a bear, an alligator, a snake—or whatever animals are relevant for your region. Most wild animals will not bother humans and are more scared of us than we are of them. We just need to leave them alone, which is a good thing to teach your child as well.

IN A NUTSHELL

Although letting kids take risks may be scary for parents and even children at first, it is an essential part of growing up. Taking risks allows children to overcome physical challenges and strengthens their senses at the same time. These benefits ultimately make them safer and more resilient in the long run. Risky play also allows children to overcome fears and anxiety and builds strong character. Children need opportunities to fail and make mistakes in order to become more confident and capable when facing future life challenges.

CHAPTER 6

What's Wrong with the Playgrounds and Indoor Play Spaces of Today?

Things are drastically changing on the play scene for children. The metal playgrounds of the past that towered over children and offered what seemed like impossible challenges have been replaced with simpler, brightly colored plastic playsets that do little to inspire growing children. At the same time, as outdoor play becomes rare, indoor play areas are all the rage.

In this chapter I discuss playgrounds in detail: how they have changed, the effects these changes have on child development, and what to look for in a good playground. I will also talk about indoor play spaces, why the outdoors offers a sensory advantage over these spaces, and what to look for in play spaces when going outdoors isn't an option.

THE PLAYGROUND DILEMMA

By the time my daughters were five years old they had outgrown most of the playgrounds in our area. If I brought them to a playground, they tried the equipment for a few minutes before growing tired of playing with it. They would instead take to the field beside the playground, starting their own games of pretend play and making forts out of sticks lying around. I started to wonder why I had brought them to a playground in the first place. A playground should inspire and challenge children, not bore them.

Playgrounds have changed since the early 1980s, when I grew up. Everything is closer to the ground. Slides are shorter, swing spans are smaller, and equipment deemed "too dangerous" has been replaced by mediocre equipment that doesn't provide the same sensory experience. These changes got me wondering about how playgrounds have evolved over the years, the reasons behind these changes, and how this evolution impacts child development.

Playgrounds of the Past

I recently found a picture of a group of schoolboys at a Dallas, Texas, playground in 1900. A great number of the boys, who looked to be between eight and ten years old, were perched, looking proud as peacocks, on top of square metal piping of the playground equipment that rose above the ground a good twenty feet! Up until twenty to thirty years ago, it was common for playground equipment to be made of metal. The boys just sat there with their hands in their laps, dangling on one bar high in the air with perfect balance. Working with countless children with balance and body awareness problems, this picture truly amazed me. To think that children back in the early 1900s had this type of stellar balance really baffled me.

Other boys were shimmying up the twenty feet of metal piping using only their upper body and core strength. Again, I was in complete awe. I have lots of "typical" kids who come to my camp program and can't even hold on to the rope swing long enough to actually use it without falling—never mind shimmying up a metal pole without any help. These kids of the past were strong!

Playgrounds started to spread across the United States in the early twentieth century with the purposes of getting children off the streets and providing a safe place to play within walking distance of their homes. Playgrounds were especially important for children who lived in the city and didn't have access to natural areas of play. Recognizing the need for public playgrounds to keep children out of danger and away from crime, former President Theodore Roosevelt stated in a speech in 1907, "City streets are unsatisfactory playgrounds for children" (Theodore Roosevelt Association n. d.).

President Roosevelt acknowledged that "play is a fundamental need" and that even older children needed to challenge their bodies. "Playgrounds should be provided for every child as much as schools. This means that they must be distributed over the cities in such a way as to be within walking distance of every boy and girl, as most children can not afford to pay carfare" (Theodore Roosevelt Association n. d.).

Playgrounds of the early to middle 1900s typically included climbing structures that were between ten and thirty feet tall, monkey bars, merry-go-rounds, twenty-to-thirty-foot-long stainless-steel slides, swings with a fifteen-foot span, and teeter-totters. They even had equipment such as the "witch's hat," a circular apparatus that children held onto and ran as fast as they could until their feet left the ground and their bodies were horizontal. Spinning in circles, the children held on for dear life.

By the 1970s and 1980s, much of this equipment was old and needed to be replaced. Instead of replacing it with new equipment of similar caliber, schools and government entities chose to use "safer" versions. In came plastic, brightly colored, and ultrasafe equipment. In came the wood chips and rubber matting. Out went the thrill of provoking, challenging playground equipment (Tierney 2011).

Rise of the Regulations

The early 1980s ushered in the "safety first" era. Restrictions became tighter and tighter due to parental concerns and the fear of lawsuits. In 1978, a toddler named Frank Nelson made his way to the top of a twelve-foot-tall tornado slide in Chicago, with his mother a few steps behind him. The boy never made it down the slide. He fell through the gap between the handrail and the steps and landed on his head, suffering a severe brain injury. A year later, his parents sued the Chicago Park District and the companies that manufactured and installed the slide. This was just one of a number of lawsuits that fueled the changes to potentially dangerous playground equipment (Rosin 2014).

In 1981, the US Consumer Product Safety Commission published the first *Public Playground Safety Handbook*. It was meant to provide guidelines to keep children safe, not requirements. It was intended to persuade manufacturers and those creating playgrounds to take a closer look at playground designs and to become aware of sharp angles, openings, and other hazards. However, people started using these guidelines for litigation purposes. Insurance premiums went through the roof (Rosin 2014). Fearing lawsuits from parents, towns starting taking away playground equipment that could potentially cause *any* harm, such as merry-go-rounds and teeter-totters, and replaced them with much simpler, unstimulating designs.

In an attempt to make playgrounds safer, we have done the extreme. We've created equipment that no longer challenges or stimulates children in ways that support healthy child development. Joe Frost, an original crusader for playground safety, has even admitted that we've gone too far with newer playground designs. He says that adults mistakenly believe that children must be sheltered from all potential risks of injury. He further states, "Reasonable risks are essential for children's healthy development" (Rosin 2014).

Change in Playground Equipment

This new era of safety obsession has drastically changed the look and feel of playgrounds. The following sections dive a little deeper into the sensory implications of taking away some of the older, challenging equipment of the past and replacing it with ultrasafe alternatives.

MERRY-GO-ROUNDS

One of my favorite pieces of playground equipment is the merry-go-round. As a child, it was the most fun! A bunch of my friends and I would climb aboard and hold on tight while one child ran circles around us, keeping us spinning. As the merry-go-round went faster, we clenched our little fists tighter. It took real strength and courage to go on a merry-go-round. As a therapist, I believe it is one of the most powerful pieces of therapeutic playground equipment around.

The merry-go-round creates a centrifugal force in the inner ear. As children spin in circles their *utricles*, which are fluid-filled cavities in the inner ear containing hair cells, experience maximum activation. The utricles send messages to the brain about the orientation of the head. The therapeutic effects of this vestibular stimulation are "centering, grounding, and sustained attention to task" (Kawar and Frick 2005). In other words, the motion of a merry-go-round promotes a calm, alert state in children while improving attention. This is exactly the opposite of what many teachers are seeing in schools! Kids are far from being calm, and many are not paying attention in the classroom.

In the early to middle 1900s, merry-go-rounds were common on playgrounds. Children with access to a merry-go-round were able to ride this thrilling piece of equipment on a daily basis. In essence, the children received rapid vestibular input, developing a strong framework for sensory integration and a capable balance system. Due to our preoccupation with safety in recent years, merry-go-rounds have been deemed unsafe and are now becoming extremely hard to find. Not only will children miss out on the thrill of using this equipment, they also won't be receiving valuable vestibular input on a day-to-day basis.

Instead of portraying the merry-go-round as a dangerous piece of equipment never to be used again, we should consider taking a step back and viewing it as an important tool for helping to prepare children for learning in the classroom. Just like every tool, we simply need to show children how to use it properly.

You may be wondering what to do with this new information. Parks are reporting that even if they wanted to replace their old and rusting merry-go-round with a new one, they can't. Most manufacturers are not building them anymore. However, there is hope. Many playground companies are making modern versions of the merry-go-round called the Supernova or the OmniSpin. They are not as challenging or therapeutic as the merry-go-round, but they do replicate most of the sensory input. Another option is to make your own merry-go-round at home or request that your town or school construct one. The directions and supplies are really quite doable and can be found online.

SWINGS AND SLIDES

The merry-go-round isn't the only piece of equipment that has changed in recent years. The chains of swings have also gotten shorter. Changing the length of swings is a physics lesson gone wrong. Doing so changes the amount of force and vestibular input children experience when swinging. Their neurological system is simply going to receive less sensory input. Again, what we think of as "safer" for children is actually causing them to become unsafe. Without daily, vigorous vestibular input, this system is going to be weak and underdeveloped, unable to do its job supporting higher-level skills such as attention, regulation, and learning.

Not only are swings shorter, but kids are also restricted in what they are allowed to do on swings. For occupational therapy homework, I recently told two of my clients, one in kindergarten and one in third grade, to spin on their swings at school to get nice rotary movement to improve body awareness. They both reported back the following week that they were not allowed to spin on the swings at school. They lived in separate towns and went to separate schools. "What?" I asked, bewildered. "You can't *spin* on your swings at school?" The third grader said, "No, my teacher told me it was too dangerous."

Activities we now consider dangerous many adults recollect as some of their fondest memories. I still remember walking up to the "tall" metal swings as a young child. They seemed to tower over me and presented a challenge, a conquest. My friend and I started by pumping. Side by side we would smile at each other as we tried to pump in unison. Eventually, we'd find ourselves pumping out of sync and giggling at the thought of this. It took us each a few minutes of pumping to get our swing to its outer limits...and naturally, we'd test those limits. We'd pump until we felt a little hitch and a quick drop high up in the air. It was thrilling.

At other times, we would spin in our swings in quick circles, going faster and faster. I even recall trying to jump off the swings to see how far we could get. We felt like we were flying, unstoppable. This brought about great feelings of confidence and joy. I never just went *swinging*—I was constantly seeking a challenge. There had to be some element of "danger" in order to keep my interest. I'd avoid the swings altogether if no risk was available. Without the potential to thrill or provide challenge, I'm afraid swings are losing their appeal to the children of today.

Like swings, slides have also gotten shorter. This simply means less vestibular input in a linear (back and forth) direction. Again, receiving vestibular input on a regular basis, and getting enough of it, is essential to developing a strong balance system. Children are also often told not to climb up slides. It makes sense for children not to climb slides if there are children coming down. However, letting kids climb up slides when no one's around is a great way to help them strengthen their upper body and improve motor control. Going down the slide on the belly, feet first, is also a great activity for kids. By changing directions and the position of the head, children activate different parts of the brain.

Instead of reflexively saying no when something your child is doing feels like it's against "playground etiquette," try looking at the playground equipment with fresh eyes. Kids have few or no preconceived notions about how to use the equipment; as far as they know, going up a slide is as much a part of its purpose as going down. Try adopting this open-minded thinking yourself, and be open to having your children use the equipment in a variety of different ways. Doing so will expand their sensory experience and allow children to assess their comfort level by using their body in new ways.

JUNGLE GYMS

The monkey bars are probably the closest thing we have to a jungle gym these days, as well as some miniature jungle gyms that are close to the ground. Both pale in comparison to the jungle gyms of the past—multilayered, towering works of art that posed a real challenge. Children of past generations took pride in going upside down and yelling to adults or peers, "Look! No hands!" It was a fun trick that only the bravest would attempt.

Going upside down moves the fluid in the inner ears and helps improve spatial awareness. Climbing heights advances children's motor skills and promotes a sense of confidence and accomplishment. Nowadays getting to the top of a jungle gym doesn't present children with much of a challenge because they are much closer to the ground. Children are also restricted from how they use monkey bars and jungle gyms. Going upside down is not acceptable. One teacher also reports, "We can't even let the kids climb across the monkey bars. There is always the danger they could fall."

In order to foster vestibular health in children, consider letting them practice going upside down at home. If they have monkey bars or even a hanging bar on their swing set, most kids will naturally attempt to go upside down without any coaching from an adult. Sometimes all it takes is watching a friend go upside down for a child to try to replicate this skill. The key is to let them.

TEETER-TOTTERS

The teeter-totter is also becoming a thing of the past. Not only did it pose an exciting element of risk, but you also got to use the teeter-totter with another child, providing opportunities for eye contact and social connection. It was a team effort, requiring kids to take turns in order to make this simple but fun piece of equipment go up and down. Although simple in design, the teeter-totter helped children work on enhancing their gross motor skills. It required good body timing, balance, core strength, and attention in order to operate effectively.

Now manufacturers are creating modern cousins of the teeter-totter that are shorter and spring-loaded in order to prevent children from going too high. These new versions don't provide the same sensory input or challenge. However, just like the merry-go-round, the teeter-totter is actually easy to make and instructions can be found online.

Merry-go-rounds, tall swings and slides, giant jungle gyms, and teeter-totters all have one thing in common: they provide an essential element of challenge combined with the necessary sensory input to nurture a developing brain. Instead of taking these therapeutic tools away from children or altering them so that they lose their therapeutic value, children would benefit from us showing them how to use the equipment correctly. It's reasonable to put safety measures in place, but they should support our growing children's needs.

What to Look for in a Good Playground

Even though most playgrounds have changed considerably from just twenty to thirty years ago, there are still some great playgrounds being designed today. Numerous blogs, such as the popular *PlayGroundology* (https://playgroundology.wordpress.com), make a point of charting all the playgrounds that still provide a challenge and thrill for children all over the world. Merry-go-rounds and teeter-totters, though rare, can still be found—if you search for them. If they can't be found, they can be made. At the same time, natural playgrounds, which promote creative play and exploration, are becoming extremely popular and are growing in number. Therefore, finding a good playground is very possible. The following section offers tips on what to look for when searching for a quality playground.

NATURAL COMPONENTS

To evaluate a playground's space and potential, it is beneficial to first notice what elements of nature are available to your child. Playing in natural areas adds an element of adventure. Nature is unpredictable, and exploring nature is always an exciting experience. Natural spaces also promote pretend play. Since playground equipment only serves a few functions, it often leaves little to the imagination. Having access to movable natural items, such as branches, rocks, sticks, leaves, and pinecones, adds a new element to play. Children will often pick up branches to build forts and pretend stores. Maybe the leaves become favorite foods. Rocks may be piled high to add to buildings or be used to create fairy houses. Natural playgrounds tend to inspire creative play in children because there are endless play opportunities.

Also, as we learned in chapter four, nature provides the ultimate sensory experience. Features such as a trickling brook and surrounding trees not only enrich the play experience, but the sound of the water and presence of nature provide calming sensory input, helping to relax and ground children. Here are other things to consider when selecting a playground:

- Surrounding wooded area, canyon, beach, or other major natural feature for exploring

- An area for water play

- A field or open area in which to play games and run around

- Slices of trees to balance on

- Gardens that can be watered or flowers that can be smelled

- Loose things that can be moved around and used to build, such as sticks, stones, and logs

- Trees to climb

- Rope or tire swings

- Piles of dirt or a sand pit to play in

- Climbing nets

- Small and large rocks to climb on and jump from

- Slides that fit into the side of a hill

SPACE TO MOVE

Simply having space to run and play is another key factor in finding the ultimate playground. The more space children have, the more privacy and ability to roam they have. It is important to be able to see the children from a distance, but when children are given their own space to play, they are more likely to engage in creative play with peers.

With more space, children have the ability to move around their environment with ease. Some kids can play a game of pickup soccer or kickball in one part of the playground while others go off in small groups to dive into pretend play. Also, giving children a bigger area to play allows for important time away from the adult world, providing them an opportunity to relax and move at their own pace, unhindered by adult fears. Here are some types of space to consider:

- A lush field of grass to run through

- A dirt area to play in (and get muddy in if kids want to)

- Bike paths or walking trails

- Large and lengthy tunnels to crawl through

- A beach or body of water to explore

- Hills to climb up or roll down

- Trees or a wooded area to explore

- A labyrinth made with rocks or shrubs to wander through

- Large rock formations to climb

- A marshy area to explore

EASY ON THE COLORS

A friend of mine once described the modern-day, brightly colored plastic playgrounds as something closely resembling a fast-food chain. The colors switch from bright red to lively blue to neon yellow. Every part of the equipment is vibrantly color coded, as if children need this extra visual hint that one piece of equipment has ended and a new one has begun. To an adult's eye, these playgrounds are visually stimulating. To a child's eye, they can possibly be overwhelming and downright distracting, causing overstimulation and a hyperactive state of arousal. Parents and teachers often observe that children get really loud and a bit disorganized when playing on this type of equipment—behavior that occupational therapists attribute to visual overstimulation.

Instead of color-coded plastic playground equipment, all connected together in one giant mass of trivial chaos, children really benefit from equipment that is simpler in design:

- Wooden or metal equipment

- Neutral colors to prevent visual overstimulation

- Logs and tree stumps integrated in the playground design

- Gardens or other natural landscape designs interspersed throughout the playground (in chapter four we learned that simply looking at nature is calming)

- Woods or water bordering the playground, which have a calming and grounding effect

SIMPLE BUT CHALLENGING EQUIPMENT

Look for playgrounds that have equipment that challenges children and provides a variety of sensory experiences. Think "old school" when looking at equipment. Merry-go-rounds, teeter-totters, tall climbing towers, swings and slides that reach great heights make excellent additions to any playground. Look for playgrounds that spread out their equipment instead of having it all in one location; this inspires children to explore and not focus solely on the equipment.

Also, consider places that offer just a few pieces of playground equipment. Too much equipment can distract children from playing creatively. With just a sampling of equipment, children who need to challenge their strength and balance can go to these stations. Otherwise, children can run around, make up their own games, create and build forts, and play "pretend" in the natural spaces. Here's what to look for:

- Fewer pieces of equipment, to avoid overwhelming children

- Tall swings and slides

- Equipment that spins, such as merry-go-rounds

- Equipment that challenges balance

- Equipment that requires two or more children to operate, such as a teeter-totter

- Equipment that encourages climbing

INDOOR PLAY SPACES

Indoor play spaces and birthday party locations are becoming extremely popular, especially in the colder months when many people shy away from going outside. While indoor play spaces can be fun and serve a purpose, such as getting out of the extreme cold, it is best to use these spaces as an occasional treat versus a replacement for the very therapeutic outdoors. As with technology, using indoor play spaces too often can replace much-needed time outdoors.

Many indoor play spaces are specifically designed to entertain children. A trampoline or bouncy house is to jump on. A ball pit is to jump into. A tunnel is for crawling through, and so on. Although children are at least getting movement in these spaces, they do little to inspire hours of pretend play. As we learned in chapter four, bright colors and loud noises can also overstimulate children and put them in a hyperactive state of mind. Children often tend to run around indoor play spaces, trying out piece after piece of equipment in a disjointed manner. It is rare for children to sit down and pretend to have a tea party or to pretend they are running away from an imaginary dragon in an indoor gym. Instead, they say things like "Let's go to the race cars now," or "Let's go jump in the pit!"

While indoor spaces may be marketed as encouraging sensory development, this is only half the truth. Yes, these spaces provide sensory input. Everywhere you go and wherever you are, you are always receiving sensory input. However, some sensory input is better for your child's development than other input. Loud noises and bright colors provide high-energy stimuli. They put children into a heightened state of arousal—not a healthy state to stay in. On the other hand, going outdoors provides a calming and organizing environment that lays the foundation for learning and creativity. Whereas indoor environments can overwhelm and overstimulate, natural settings inspire, revive, and restore.

Going to an indoor play space is like giving your children soda. Is it going to kill them if they go from time to time? Probably not. In fact, indoor spaces offer a new and fun experience. Is it something you want them to have every day? Definitely not. It goes back to my motto of "everything in moderation." Indoor play is fine as long as it doesn't replace or even outweigh time spent outdoors. We see problems with sensory development when children spend most of their time indoors.

There are times when the outdoors may not be accessible or when the weather is frigid. Or maybe you want to try an indoor space just for the fun of it! In this case, do some research and be particular about where you bring your child based on the environment and your child's needs. Below are some suggestions for what to look for in a quality indoor play space.

- **Museums**—Children's museums and science museums are fascinating indoor spaces that often provide both space to move around and displays that inspire children to think creatively. Although they may have crowds, they often lack bright colors and plastic-filled rooms. Instead, they offer interactive exhibits, giving children hands-on experiences they may not encounter otherwise. Everything from seeing chicks hatch in an incubator to driving a submarine simulator can offer novel experiences for children.

- **Water parks**—These aqua places are not only a blast for children of all ages but offer a great vestibular and touch experience. Going down long slides at different speeds gives children essential vestibular input. Splashing water in the face and all over the limbs improves tolerance for touch experiences. If the water park is outdoors, even better. The outdoors adds the elements of sun, wind, and more for a robust sensory experience.

- **Public atriums**—These wide-open spaces often include indoor gardens. Although children cannot really let go and run around in these open spaces, atriums often offer a natural scene. As we learned in chapter four, simply looking at nature is calming and grounding for young children. This is a good place for quiet play and reading stories.

- **Playgrounds**—If your child really needs to just let loose, indoor playgrounds may be what you need when going outdoors is not an option. Look for playgrounds that challenge and inspire. Playgrounds that offer loose (movable) parts are best because kids can design and create their own play spaces. Also, look for high structures to climb and equipment that gets children moving in all different directions. I also suggest going where there aren't too many colors or too many people. Places with neutral tones and fewer people provide children with a calm but challenging environment to explore.

- **Rock-climbing gyms**—It's ideal to go bouldering or rock climbing outdoors, where children have to use their imagination and problem-solving skills at a more complex level. However, when the outdoors is inaccessible, indoor gyms can be an acceptable alternative. Rock climbing improves children's motor skills, problem-solving abilities, body awareness, and overall strength. Children also learn patience, perseverance, and control, and they develop confidence as they scale the walls.

- **Aquariums**—Aquariums expose children to a world they may not normally get to encounter. Visually stimulating but not overwhelming, children get to see deep-sea creatures of all sizes, shapes, and colors. Some aquariums offer interactive touch tanks, where children get to excite the tactile sense by feeling different sea animals, such as sea urchins and slippery stingrays.

- **Swimming pools**—Whether indoors or outdoors, swimming offers great therapeutic value. Water offers buoyancy properties that provide a sense of weightlessness and reduce impact on the joints and muscles, allowing children to move freely while still getting the resistance needed to improve overall strength. Also, as children move through the water it touches all parts of the body and gives great tactile feedback. Swimming is also great fun, and children can experiment with jumping into the water, spinning in circles underwater, and holding their breath.

- **Art or science (tinkering) labs**—Exploring open-ended art and tinkering projects always stretches the mind and gets children thinking creatively. When children are given beautiful art materials and tinkering objects with no specific purpose, they start with a clean slate and an open mind. The design options are endless, and children learn the critical skills of conceptualizing an idea, executing that idea, and then evaluating the end results. This stimulates their imagination, creativity, problem-solving skills, and self-evaluation skills.

- **Petting zoos**—One-on-one interactions with live animals are always engaging sensory experiences. Feeding animals out of the palm of the hand provides children with great tactile feedback. The noises of animals help children to recall this experience and form important neural connections and memories.

IN A NUTSHELL

Times have changed: children are playing far less per day than children twenty and thirty years ago. Play has changed: children don't play as often with children of other ages, and rarely do they play without a parent shadowing their every move or potential fall. Play spaces have changed: playground equipment is less challenging and stimulating, and a lot of it is being moved indoors. It is more important than ever that we offer children daily outdoor play to support healthy, meaningful sensory experiences. Natural outdoor settings offer the most open-ended, sensory-calming environments for children, inspiring hours of pretend play and creativity that aid in the proper development of their senses.

However, when a natural setting isn't available or you want to change things up, playgrounds and indoor play spaces can also offer great experiences, if chosen carefully. Look for places that stimulate without overwhelming children. Look for equipment or exhibits that challenge them physically in different ways and foster creativity. Remember that these play areas are great as a treat but should not replace or take away from daily outdoor experiences.

CHAPTER 7

Rethinking Recess and the Classroom

There are only two rules at TimberNook: (1) children have to be able to see an adult at all times, and (2) children need to show respect to the other children and adults. That's it. Everything else is fair game. Kids can climb trees, yell, run, jump, go barefoot, use tools, get dirty, create their own societies, and build with whatever they find out in the woods. This often surprises adults. "Doesn't this create chaos out there?" they ask. For many of them, the classic book *Lord of the Flies* quickly comes to mind. In reality, we see less testing of limits and fewer behavioral issues than we observe in the local elementary schools—inside buildings with a plethora of rules.

That children seem calmer, more focused, less aggressive, and truly passionate during their play experiences at TimberNook doesn't surprise just parents. It also stuns teachers, school administrators, and even physical education instructors, all of whom have openly wondered whether what is done at our camp is translatable to recess and the classroom. It is.

Remember Bruce McLachlan from Swanson Elementary School in New Zealand in chapter five? Bruce got rid of his rules at recess time and saw a drastic decline in the number of children "misbehaving," as well as an increase in concentration and engagement in learning when kids returned to the classroom. We talked about the similarities between his unconventional recess and TimberNook, and we discovered key overlapping factors that make our programs successful and far from chaotic. This chapter talks about these important qualities as well as how to implement them during recess, in classroom settings, and in day care environments to enhance child development and engage children in the learning process at a deeper level.

RETHINKING RECESS

I still remember recess like it was yesterday. Our play area backed up to a wooded lot and we were allowed in the woods as long as we could still see the teachers. I remember playing amongst the trees with my friends, pretending to be a princess or some other fairy tale character. Once in a while we tried the jungle gym or would jump off the swings. Sometimes my friend Crystal and I would even play a game of pickup soccer with the boys. Other times, we were showing off the round-off back handsprings we had learned in gymnastics.

We played for a full hour before it was time to go back inside. In the classroom again, we were fully energized and excited to sing songs, participate in a science project, or engage in whatever learning experience the teachers had in store for us. I *loved* recess and I *loved* school. Times are changing. My own daughters started saying they hated school at the young age of five. What has changed? Recess and the time available to be a child are slowly disappearing.

Up until the late 1980s, it was not uncommon to have a full-hour recess session and another one or two shorter recess sessions (Nussbaum 2006). However, in the 1990s improving achievement scores became a critical issue for schools and legislators, and it seemed logical that increasing time in instruction would improve success with these tests. With limited hours during the day, recess became the most rational activity to reduce, despite a growing body of research that demonstrated the importance of recess for child development and school achievement (Pellegrini and Bohn-Gettler 2013).

By the late 1990s, 40 percent of school districts in the United States had either reduced or eliminated recess for good (Zygmunt-Fillwalk and Bilello 2005). When the No Child Left Behind Act passed in 2001, recess times continued to be cut as there was growing pressure to meet new standards of achievement (Pellegrini and Bohn-Gettler 2013). Now children are lucky if they have one fifteen- or twenty-minute recess session in a six-hour period of time!

Recess Can Make Your Child a Better Student

As we learned in chapter three, it is important to allow children hours every day to move and play in order to support attention in the classroom and healthy sensory integration. Recess, therefore, is an ideal opportunity to expose children to quality play experiences. It may not be the beneficial hour of yesteryear, but any movement break for children is better than nothing. Recess nourishes the growing mind and body. It provides time to move and time to engage in active free play—both essential for fostering healthy child development. The following are just a few of the ways recess benefits child development and academic success.

- **Combats obesity**—In a time when more children are becoming obese, extending recess times may be an important step toward getting children to become more active. When children are free to play at recess, studies have found that at least 60 percent of them engage in physically active social play (Pellegrini 2009). In fact, research shows that children may even engage in higher levels of active play during recess compared to during organized physical education class (Kraft 1989). In other words, recess—especially outdoors—encourages physical activity and physical fitness.

- **Improves behavior**—Studies indicate that recess improves children's ability to concentrate in the classroom, reduces the need to fidget, and improves overall behavior (Jarrett et al. 1998; Pellegrini and Bohn-Gettler 2013). That's because ample movement throughout the day develops a strong vestibular system, which supports attention and engagement in the classroom and also helps regulate emotions in children. Teachers have observed that the longer the recess, the more attentive and engaged children are afterward in the classroom.

- **Develops social skills**—Recess provides an important avenue for free play with a variety of other children. As children play with each other, they learn how to negotiate, take turns, communicate effectively, listen, assert their needs, deal with conflicts, create and follow rules, and practice leadership skills. Adults can try to teach children these skills in social-skills groups and through role-playing. However, playing with friends offers real-world opportunities to practice and learn these skills at a much deeper level.

- **Gets the brain working**—Without recess or a break from academic work, the brain becomes less efficient and less attentive as time elapses (Jensen 1998). Research suggests that breaks need to be more drastic than just walking to another classroom or switching subjects. In fact, children really need to participate in play and movement in order to stay engaged and attentive with their studies (Ginsburg 2007).

- **Reduces stress**—The National Association for the Education of Young Children recommends free play as a developmentally appropriate way to reduce stress in children (Pica 2010). Recess gives children the opportunity to engage in free play away from the adult-driven rules and regulations standard in many schools today. Instead, they get to choose their play schemes, pick with whom they are going to play, and partake in making up their own rules to games. It is also their opportunity to let loose, yell, run, and even say no if they want to. Recess offers them independence and the chance to let off steam, which is especially important during long school days.

It is hard to ignore the growing body of research pointing to the many benefits of recess for growing children. Not only is the amount of time children spend in recess critical, but the environment and the way recess is conducted are also important to note. The following section talks about the significance of offering recess as an enriching play experience.

Ways to Make Recess a Play Experience

I've heard many people talk about offering recess as a more structured time, in which adults coordinate the activities and get the children moving. Although the intentions are good, children really need this time away from adult instruction to think on their own, move the way their body needs, and have an opportunity to dive into their imaginations. Recess is best left untouched by the adult world. The following are four simple guidelines that will enhance the recess experience for children.

EXTEND THE TIME

Not only do children need plenty of time to move and activate their senses in order to support healthy sensory integration, but they also need time to dive into deep play. I've observed over and over that it takes an average of forty-five minutes for children to find out who they are going to play with, decide what they are going to play, and finally come up with a play scheme. With a twenty-minute recess session, children are often just figuring out whom they are going to play with when the bell signals the end of recess. Twenty minutes is not enough time to engage in the type of play that challenges the mind and body.

FEWER RULES

We often think that more rules will make children comply, listen, and behave, when in reality the opposite is often the truth. The Swanson Elementary School in New Zealand did a study with the University of Auckland to see what happens when rules are removed from recess. The principal noted that they saw a significant decrease in unruly behaviors and no longer needed a time-out area or as many teachers patrolling the play area. The children were too busy playing to act out. He says, "The kids were motivated, busy and engaged. In my experience, the time children get into trouble is when they are not busy, motivated and engaged. It's during that time they bully other kids, graffiti or wreck things around the school" (Saul 2014).

The same is true at TimberNook. We find that with fewer rules there is less testing of boundaries. Children are so motivated to engage in play with their peers out in the woods that they focus mostly on their play experience. Also, by having fewer adult-driven rules and allowing children to help us come up with the rules, they internalize the rules and are more likely to follow them. In general, letting children take risks, as we learned in chapter five, helps them become more coordinated, become better problem solvers, and to be safer in the long run.

LOOSE PARTS

"Loose parts" are becoming popular in the world of early childhood education. The term and concept were first proposed back in the 1970s by architect Simon Nicholson, who believed that loose parts (things that can be easily moved) in the environment inspire creativity in children (Kable 2010).

Loose parts are basically materials that children can use to design, create, move, and play with. Larger loose parts may be wooden planks, rocks, bricks, tires, large sticks, hay, and old hoses. I've seen children use all of these objects to create forts, societies, stores, traps, and boats.

Smaller loose parts may be shells, pea pods, acorns, pine-cones, baskets, masks, transparent fabric or blankets, tarps, clothespins, pieces of rope, and tape. Children use these objects during the building process and to further refine their imaginative play. For instance, pinecones may become the children's rate of currency. Acorns may become the food they cook. Rope and tape may be used when building a pulley system or creating an elaborate fort.

Large loose parts such as logs and bricks should be kept on the ground and organized in piles at the start of recess, especially at the beginning of the week. If they are put away in bins, the children may not see or use them. Put them in a central location as they are the foundation for creative play and building. Shears should be placed in an obvious location as well. You can drape them on a clothesline or the low-lying branches of a tree. Smaller loose parts should be placed near the larger loose parts. They can be placed in inviting baskets or pots and pans. No explanation is necessary. Children will find them on their own.

It's best not to point out the loose parts to children; this is part of the discovery process. By discovering the parts on their own, children then take ownership of the supplies provided. Adults are encouraged to step back and simply observe the children's interactions and usage of loose parts. When given the freedom, children will create play schemes we would never have dreamed of.

It is important not to offer too many loose parts at one time. Doing so can be visually overwhelming, and children may avoid using the parts altogether. The bigger loose parts that are typically used for construction purposes—logs, bricks, tree cookies (small cross-sections of tree trunks), and tires—should always be available to the children. However, rotate only a few smaller loose parts at a time. Experiment with what is popular and what gets used most often.

If children create forts and other large structures with the loose parts, ideally you should leave these for the week. Explain to the children that you'll be cleaning up the recess area once a week so that they can start the following week with a clean slate. However, if you have the luxury of letting some of the structures stay up for longer, their play schemes will get more and more elaborate as time goes on. Decide as a group what you'll clean up and what you'll leave for the children to continue their imaginative play ideas each week.

By adding loose parts to recess, you are giving children tools to experiment with and to incorporate into their worlds of play.

FREE TO GET DIRTY

Children should be free to get messy and dirty—another commonality between the Swanson Elementary School and TimberNook. Children play in mud puddles, make mud cakes and soup, romp in mud slides, and then get cleaned up before going back into the classroom or back home. Children simply need a towel, a change of clothes, and time to get changed.

The principal at the Swanson Elementary School says that letting kids get dirty was one of his biggest barriers to overcome with teachers. The teachers were not keen on having children come into the classroom dirty, nor did they like sacrificing time to allow the children to change. However, once they put a system in place, the kids were totally independent and responsible for changing quickly and efficiently in fewer than ten minutes after recess.

By making just these few minor adjustments to recess, adults are likely to see drastic changes in their children's learning abilities, sensory abilities, and behavior. All it takes is time, creating a space that enhances creativity and sensory play, and letting go of having total control over what we allow our children to do. The rest will take care of itself.

RETHINKING THE CLASSROOM

A typical classroom setting includes chairs lined up in rows, colorful printouts decorating all four walls, a rug for circle time, and a desk for the teacher. This is pretty standard. However, this environment does little to stimulate creative play experiences—the type of play that gets children exploring, creating, and problem solving. In order to foster these skills, challenge yourself to rethink traditional classroom design and operation. The following are some strategies for changing the classroom.

161

Keep Things Visually Simple

The Waldorf schools are famous for using natural supplies, such as wool, cotton, pinecones, and fabric, in their classrooms. Everything is kept visually appealing, but simple in design. A child may play with shears, a simple rag doll, or wooden blocks. The idea with the classroom environment is to keep supplies basic and open-ended to inspire kids to use their imagination to give objects meaning and life. Not only does offering basic natural materials in the classroom prompt children to use their imagination, but by keeping things visually simple, children are less likely to get overwhelmed or to become overstimulated by a multitude of colors and plastic objects.

Get Moving in a Meaningful Way

Children don't need to sit in order to learn. In fact, the opposite may be true. Most children enter kindergarten as kinesthetic (hands-on) learners, moving and touching everything as they learn. By second or third grade, some students become visual learners. During late elementary school, some become auditory learners. Yet many adults maintain kinesthetic strengths throughout their life (Dunn and Dunn 1993). Kinesthetic learners are most successful when they are totally engaged in a learning experience. They acquire information the fastest when they are doing science experiments, putting on a play, taking a field trip, exploring the outdoors, designing, dancing, or performing any type of active activity.

In order to appeal to all types of learners, it is best to incorporate movement into the actual learning experiences. Not only will longer recess sessions and play after school help children develop strong minds and bodies, but using whole-body learning experiences also enhances the learning process.

Sitting on a bouncy ball or a wiggle cushion is a method occupational therapists often prescribe to help children focus in the classroom. Although this is a popular method to get children (especially children with attention deficit/hyperactivity disorder) to pay attention for a few minutes, recent research indicates that doing so may also distract most children and actually hinder their concentration (Reddy 2015). Therefore, instead of focusing solely on trying to incorporate simple movement techniques (such as fidgeting or using therapy balls) in the classroom, we may be better off getting children up and engaged in active, meaningful learning experiences. The following are some strategies to get you started.

SIT AND ATTEND FOR BRIEF PERIODS OF TIME

Genuine *external* attention, such as listening to a teacher lecture or explain a new concept, can be sustained for only a short time, generally ten minutes or less (Jensen 1998). Try keeping instruction to a minimum, and then follow it with movement to help children process the new information. For instance, children can do a ball-toss activity to practice new math facts or go for a brief walk to discuss a new concept.

CHANGE POSITIONS OFTEN

Have children experience learning in different positions throughout the day. Every ten to fifteen minutes, have them either move or change positions in order to prevent constant gravitational loads and pressures on specific joints. For example, maybe they sit in a chair for ten to fifteen minutes while completing a writing task. For the next ten to fifteen minutes, have them stand or squat in a circle to discuss or share ideas about their writing project. You could have them spend the next few minutes lying on their bellies listening to a story.

THINK BEYOND THE CHAIR

For anything that doesn't need to be done sitting in a chair, change the activity to incorporate different ways of using the body. Get creative. We certainly don't need to sit in a chair in order to learn. In fact, it is unnatural to expect children to sit in a confined spot for most of the day. Let's look at a few examples to help you think in different ways.

- A writing task can be done on the floor, with two children sitting back-to-back. Let's say both children write a sentence on their paper. Then they can switch papers and add a sentence to their partner's story. They can continue until there are two separate stories. By sitting on the ground, the children have the luxury of sitting in whatever position is most comfortable. For instance, one child may extend his legs, while the other child may choose to cross her legs.

- Art projects can be done with kids standing at easels or work that is taped to the wall.

- Math can be learned by playing movement games in the classroom. For example, you can reinforce geometry concepts by having children make the shapes with their bodies. Children can play hopscotch using addition and subtraction concepts. Singing songs about numbers while moving in a circle or tossing a ball around is another simple idea.

GET UP AND DANCE

If you don't have time to go outside and take a walk or explore, try dancing. Dancing is a meaningful and fun experience for children. It gets them moving in all different directions and challenges their core and balance at the same time, all of which are important for stimulating and igniting the brain to pay attention. The dance can be as simple as the freeze dance, in which children dance in place until the music stops and they must freeze in position. Or children can help create the dance routine.

You can also incorporate spinning and going upside down in the dances to make changes to the vestibular system, which will improve balance, attention, and coordination if done on a daily basis.

PROJECT-BASED LEARNING EXPERIENCES

Project-based learning is considered an alternative to paper-based learning. Children explore real-world problems and challenges and acquire a deeper knowledge by experiencing their lessons firsthand. Examples include creating a store in the classroom, putting on skits, working on design or science projects, painting life-sized murals, or choreographing a dance.

A more specific example is creating a restaurant. Children can take turns being the host, waiters or waitresses, chef, and so on. The people attending the restaurant can have a real sit-down meal experience. Kids can sew their own aprons and costumes for the event, interview people about the type of food they would like to be served (for example, fish tacos and bean burritos if they decide to do a Mexican restaurant), make the decorations, and make the food. This can become a multifaceted learning experience that not only gets the children moving and engaged but teaches them all aspects of the endeavor. This restaurant experience can be further enhanced if done outdoors, because changes in the wind, sun, and temperature engage multiple senses at the same time.

Nature in the Classroom

We learned in chapter four that just looking at nature soothes children and helps with overall mood. Simply having plants in the classroom can have a calming effect on children. Better yet, children can grow their own plants from seeds. Another way to bring nature indoors is to have baskets of pinecones, acorns, chestnuts, and small slices of trees placed out for children to build and create with. Classroom pets provide opportunities for children to learn how to care for other living things, and they can be therapeutic.

You can also get really creative and design your classroom with nature in mind. Having giant stick-mobiles hanging from the ceiling to display children's work, large pieces of tree bark lining the walls, an indoor garden area, and a nature-weaving wall (a vertical space in the classroom where children weave pieces of nature, such as feathers and leaves, in between pieces of twine, yarn, or fabric) are just a few ways to make the classroom come alive. Hatching chicks and butterflies and making ant farms are also fun methods to get children engaged with what goes on in the natural world.

Bringing the Classroom Outdoors

One of the best ways for children to learn is to bring the classroom outdoors. The classroom could be picnic tables to sit at, a large patch of grass, or a large tree to sit under. Even using the blacktop can suffice. Learning outside provides more opportunities for risk taking, problem solving, moving the whole body, using the imagination, overcoming fears, engaging in teamwork, and tolerating and integrating new sensory experiences.

Remember the math games I suggested earlier in this chapter to get children moving? Do these outdoors instead. Draw hopscotch lines with chalk on the blacktop for number games and have kids make shapes using their bodies in the grass instead.

Children can write poetry or journal entries outdoors. In fact, being outside will not only ignite their senses and help with attention, but the natural scene in front of them may inspire unique thoughts and ideas. Go even further by creating a school garden outdoors. Plant vegetables and then use them for a cooking project. Planting and nurturing a garden, and cooking with its produce, incorporates writing, math, and problem solving while engaging the senses at the same time.

I interviewed Jessica Lahey, a middle-school teacher and author of *The Gift of Failure: How the Best Parents Learn to Let Go So Their Children Can Succeed*, on the subject of bringing kids outdoors for learning opportunities. She recalled her days teaching middle school at the Crossroads Academy in New Hampshire. She often took her whole English class for walks on the paths through the woods. She'd give them a topic to ponder and then they'd walk for ten minutes to think about it. At the end of the ten minutes, they'd huddle and discuss the topic. Then, she'd throw out another one, and they'd walk some more.

Lahey also had her students write poetry outdoors with nature as their inspiration. To work on public speaking, she had some children stand on a bridge that went over a roaring brook while the rest of the class stood on the shoreline. The children on the bridge had to learn to project their voices loud enough so the children on the bank could hear their speeches.

Jessica went on to explain that children in urban schools can get outdoors just as often as children in rural settings. They can walk to the local opera house to watch a play, walk to a park, or walk to museums. Children can bring along their writing and sketch journals and assess the world and culture around them. "Learning doesn't have to be done in a chair," she says, no matter where the school is. All it takes is creativity to move beyond

167

traditional structures and get rich memory-making learning experiences.

I've always been inspired by Finnish schools. Instead of simply learning about fish from a textbook or lecture, kids spend time knee-deep in a slow-moving brook examining fish as they swim past their feet. They also learn about the whole habitat and examine the plants the fish eat, the other creatures in the brook, and much more. Later, they dissect the fish. In one video (Thai Teachers Television 2012) on Finnish education I watched, a little boy found a smaller fish inside the bigger fish. All the kids crowded around to get a closer look. Through that hands-on experience they learned that fish do in fact eat other fish.

Even Finnish physical education classes are held outdoors, but not in the way many American schools would do so today. In another video (BBC News 2010) I watched, children were geared up to go cross-country skiing. Off they went on their own for about an hour's time. The kids were around ten years old and knew the ski route well. No adult was necessary.

Forest kindergartens are all the rage these days. A forest kindergarten can be described as a school without walls or ceilings. The children and staff spend most of their time outdoors, no matter what the weather is like. Children are encouraged to play, explore, and learn in a natural environment. Forest kindergarten teachers rave about the benefits. They see more creativity, more engagement, enhanced social skills, and advanced problem-solving skills (Neate 2013). The physical activity students engage in during free play in these schools is also linked to improved reading, math achievement, and overall general intelligence (Centers for Disease Control and Prevention 2010).

Forest kindergartens are popping up everywhere because people are starting to see the benefits of letting children spend a good portion of their learning time outdoors. Inspired by this model of teaching, one teacher in Vermont went to her principal with the idea of doing a "forest Monday." In other words, she proposed that every Monday, rain or shine, she would take the children outdoors for the entire school day to learn in the woods. He surprised her by saying, "Try it" (Hanford 2015).

Now, once a week, the teacher and children head to the woods located near their school, where they built a homesite equipped with forts and a fire pit. They do both formal and informal lessons. Kids learn to write their letters in the dirt with sticks and practice math skills by measuring nature items using a ruler. Not only do the kids learn academic skills just as well as if they were indoors, but the teacher also states that they are learning essential life skills and developing "amazing grit" (Hanford 2015).

Older children need to be outdoors as well. One way to do this is by incorporating the outdoors into the curriculum. Whether that means one whole day is spent outdoors each week or parts of each day are spent outside, time outdoors can mean a world of difference in a child's educational experience.

For some teachers, integrating the outdoors into the curriculum may mean going on a field trip once a week or once a month to a natural setting if there isn't one behind their school. However, ideally the children can simply walk to, play, and explore the woods, green space, or fields beside the school. For instance, if there is a little creek behind your school, go explore it and have the children make observations of what they see, hear, and smell. If there is a patch of dirt that receives a lot of sun, create a garden. Look at your environment for inspiration and see what can be used as an area of exploration to enhance the learning process.

RETHINKING DAY CARE

The benefits of free play and movement are even more pronounced among younger children. And, compared to those who play indoors, those who play outdoors have a wider range of developmental benefits: they tend to be more creative, learn how to regulate emotions more efficiently, use imaginative play more frequently, become interested in working with friends toward a common goal, create their own rules, and start to learn how to work out problems without needing constant reassurance from adults.

Teachers from a local child care center who trained with TimberNook's nature program and implemented more time outdoors noticed significant changes in the children they cared for, even over the course of just one week. One teacher says, "The amount of creativity, problem solving, peer conflict resolution, communication, movement, risk taking, fun, peace, and wonder that we observed in the woods is hard to share with words. I saw children grow more in one week than we often see over the course of months." They were so inspired by these changes that they started taking the children outdoors on a regular basis, allowing them to take more risks and engage in lengthier bouts of free play.

Young children process information differently than older children. They are less efficient at processing information and require more cognitive effort to complete tasks, making sustained attention a challenge. Therefore, many researchers argue that younger children have a more functional-based way of thinking that is best suited for settings that promote free play (Pellegrini and Bohn-Gettler 2013). The following sections discuss steps you can take toward offering a play-centered outdoor day-care environment.

Spend Most of Your Time Outdoors

Day care can be stressful for children. In fact, research has indicated that while day care can improve reading and math scores, it typically has a negative effect on social behavior, often leading to increased aggression, talking back, cruelty, and temper tantrums. Researchers are finding that these behavior problems are due to increased levels of stress and cortisol that increase further through the course of the day (Geoffroy et al. 2006). To combat this problem, one solution is to bring the children outdoors.

As we learned in chapter four, children thrive when they spend time outside. Playing outdoors improves the immune system, develops the senses, strengthens motor skills, inspires creativity and imagination, fosters social-emotional skills, and cultivates the foundational skills needed for later academic work. After spending hours outdoors every day, children start to acquire more resilience, stamina, and independence, and overall they are calmer and more confident than their peers (Harrison, Harrison, and McArdle 2013).

Consider a Multiage Approach

Children in hunter-gatherer cultures often took care of each other. In fact, typically the older children (sometimes only four years old) were responsible for caring for babies and younger children (Dewar 2011). According to researcher and play expert Peter Gray, both younger and older children benefit from mixed-age play (2013).

Gray believes that younger children can engage in and learn from activities that would be too complex, difficult, or dangerous for them to do with peers their own age or by themselves. They also learn from simply watching older children interact in more sophisticated play experiences. And they can receive emotional support, understanding, and care that they may not receive from same-age peers. They also gain independence, resilience, and confidence more quickly (Gray 2013).

On the other hand, older children learn leadership and nurturing skills. They get to experience being the "mature one" in the relationship. They also gain deeper understandings of concepts through the process of explaining things to younger children. Also, younger children often inspire older ones to engage in more creative and imaginative play that they might not otherwise participate in (Gray 2013). Therefore, it is highly beneficial for children of all ages to have mixed-group play experiences.

Use the Environment

The Reggio Emilia approach is an educational philosophy that centers around using a supportive and enriching environment to inspire children to guide their own learning experiences. Loris Malaguzzi developed this approach to learning that focuses heavily on the environment, which is often referred to as the "third teacher." Using the outdoor environment as the inspiration for play increases creativity as well as therapeutic potential. An easy way to enhance a day care program is to add the elements of water, dirt, and fire.

Having access to water (such as using a manual well pump or setting up a play area by a small brook or stream) and dirt will open up the children's world to new possibilities. They may fill buckets and make mud pies, take off their shoes to feel sticky mud between their toes, or even explore the brook for living creatures. They might make boats, create dams in the stream, or build bridges with friends. There are endless possibilities for play. Having group fire experiences is also an enriching process for children. They get to sample new foods (such as the taste of popcorn over an open fire), learn fire safety, and learn how to prepare and cook foods in a fun way.

Offer Loose Parts

Objects that kids can carry, play with, and build with are also a really important component in creating a play-based day care center, especially one outdoors. The open-endedness of the uses of loose parts allows for hours of creativity and imaginary play. See "Rethinking Recess" earlier in this chapter for plenty of ideas about how to safely incorporate loose parts into play outdoors.

Encourage Risk Taking

As we learned in chapter five, children develop strength, coordination, resilience, problem-solving skills, and confidence when they are allowed to take risks. If we never let them take risks, they will not acquire the motor skills needed to get to the next developmental level. We don't have to let them take drastic risks; simply have the adults step back and allow the children to practice assessing risk as much as possible on their own. Let them climb onto a tree stump and attempt jumping off. Let them ride their bike through a mud puddle. Let them try to hammer boards together. They will learn through practice how to develop new essential life skills.

IN A NUTSHELL

Free play and opportunities to move are essential for healthy child development and for fostering a lifelong love of learning. However, in order to promote more play and movement throughout the day, it's vital to rethink our current educational models. Sitting for long hours at a desk prevents a good portion of children from learning at their full potential. Instead, children benefit if we provide them daily opportunities to get outdoors, engage their senses, and learn through hands-on, exploratory play experiences. Simple changes, such as allowing children to get dirty, providing inexpensive loose parts (sticks, planks, and tires), and having adults play less of a role during free play, are likely to create lasting and powerful changes in our children's behavior and ability to learn over time.

CHAPTER 8

When Is My Baby Ready for the Outdoors?

O ne day I got a phone call from a nonprofit organization that supports parents of children with special needs.

"We love what you are doing with TimberNook!" The man on the other line said.

He was interested in holding monthly meetings at TimberNook for the parents in his group. He asked if he could tour the facility and whether we had an indoor space for the meetings. I explained that the "facility" is the woods and that we had no building except my family's home. I invited them for a tour anyway. On the day of the tour, two men came to our site dressed up in semiformal wear. We set off into the woods—not the best place for suits and ties. After viewing the outdoor classrooms, we trekked back toward the house and parking area.

"Can I see inside your home?" one of the men asked.

"Ah…" I hesitated. "Yes, but I'm not sure why you'd like to see inside my house."

"Well, the woods are nice and all, but there will be *babies* at these meetings. The parents can't possibly put them on the ground! We were wondering if we could hold the meetings in your house," he explained.

Being outdoors in nature offers children—including *babies*—an ideal environment in which to develop their sensory systems. In fact, the sooner children are exposed to the outdoors, the better their chances of avoiding many of the problems we discussed in chapter one. Even newborns greatly benefit from exposure to the outdoors. Not only do experiences in nature lay a strong foundation for later life and academic learning, they can also deepen your bond with your baby and create lasting memories.

This chapter offers simple ways to introduce babies, age zero to twelve months, to nature. I also share numerous proven benefits for babies who spend regular time outside, along with lots of advice for keeping babies out of harm's way.

Nature is extremely beneficial for young children, and it won't hurt babies and toddlers to crawl and toddle around in the woods. In fact, it's good for them! That's what I told the two men from the nonprofit organization who had reservations about putting babies on the ground during their meetings. I explained that holding the meetings outdoors would dispel misconceptions that little children and babies need to be protected from the outdoors. And best of all, it would offer a great example to parents. They'd get to witness their babies and toddlers exploring nature and maneuvering their natural environments firsthand, each week mastering a new skill and experiencing varying sensations.

As it was for the parents who eventually attended the outdoor meetings, seeing your child nourished by the natural environment is far more powerful than talking about the importance of nature in early childhood development in an indoor meeting space. Even the youngest of society can experience the joys of nature and reap its sensory benefits. So don't delay! Get out with your baby today! I'll show you how to make the most of your time outdoors in the following pages.

NEWBORNS IN NATURE
(ZERO TO SIX MONTHS)

Babies are born into this world soft, vulnerable, and with great potential to adapt to the sensory experiences around them. They already have most of their neurons; all of their senses are intact; and their bones, muscles, and ligaments are in the right places. However, in order to start effectively moving their body and making sense of the world, babies require frequent access to a rich and varied (but not overwhelming) sensory environment. Nature provides this for us. It offers a "just right" environment for babies to thrive in, providing a sense of peace while stimulating the senses to continuously adapt and evolve into more mature systems.

Taking Walks Outdoors

Infants benefit immensely, on many different levels, from being carried for walks outdoors. This activity gets them out of baby devices, ignites their senses, and provides them freedom to stretch and challenge their body in new ways.

LEAVING THE CONTAINERS BEHIND

The motto "all in moderation" is a saying I refer to often in all aspects of my work and personal life. I find it's particularly applicable to the mountain of gear parents have to transport their young children. I'm no exception. I enjoy using strollers and backpacks, which are perfect for certain types of outings. However, it can be extremely beneficial to carry your child when given the opportunity. Doing so enhances your child's sensory experience and challenges muscles.

"Baby containers" are designed fairly well these days, with bells and whistles that keep the baby comfortable and make it easy for parents to get work done. Problems arise when babies are in these devices *all* the time, which is a phenomenon happening more frequently today (see chapter two). These devices affect human development by altering the types and frequency of gravitational pressures exerted on a baby's tissues. The results of being in a baby container most of the day can range from delayed physical milestones to more permanent damage, such as changes to the shape of hip sockets and altered walking patterns (Crawford 2013).

Newborns are really soft and pliable. Their bones and muscles haven't fully formed yet. How they move and how often can alter the loads that are placed on their body, affecting bone shape and muscle development. For example, the "flat head" effect is the result of too much time spent on the back, often in containers. The head is molded into an oblique or slanted shape and becomes asymmetrical in appearance. Also, bony adaptations to the environment can happen all over the body, not just the head. Overuse of a baby container also prevents the muscle activation that is necessary to develop a strong and capable body (Crawford 2013).

SENSATIONS FROM BEING CARRIED

Newborns benefit from the sensory input that accompanies being carried outdoors. As a caregiver switches the baby from one side of her body to the other, the baby experiences varying gravitational forces. Not only does this force the baby's muscles and bones to adapt and get stronger, but it also stimulates the movement and position senses. As you vary the position of your child while you walk, you move the fluid around in his inner ear, stimulating the vestibular (balance) sense. Just like a muscle that strengthens the more it is used, our senses become more organized the more they are stimulated, as we learned in chapter two. By varying the duration of time you hold your children and the frequency with which you reposition them, you are providing them with a great deal of wonderful vestibular input.

As you carry your children around, you also give them an opportunity to push their limbs against your body, providing sensory stimulation to joints and muscles. This helps to calm children and lays the foundation for them to eventually be able to control the force and direction of their movements, which is essential for effective motor skill development.

One of the first things babies learn to do is to hold their head up and control their eye muscles. Carrying your baby up against your shoulder in a vertical position gives her the opportunity to practice keeping the eyes and neck stable while you move. The more often you carry her in this position, the more opportunities she has to master integrating the senses from the eye muscles, the movement and gravitational senses of the inner ear, and muscle sensations from the neck in order to create a clear picture of the world around her. This integration of the senses plays a very important role in survival and is essential for developing the basic head and eye control needed for later looking and listening. It continues to develop over the next few years, laying the foundation for more complex skills, such as reading and balance (Ayres 2000).

If children are constantly placed in baby carriers that totally support the head and neck, the vital integration process of the neck and eye muscles and the movement senses will not occur, leading to poor posture, disorganization of the senses, and an inability to receive adequate sensory information about position in space. Therefore, carrying babies outdoors in a manner that forces them to practice holding their head up gives them essential movement feedback from the very beginning.

Stepping Outdoors Ignites the Senses

Stepping outdoors with your baby in tow offers an array of sensations all at once. For instance, facing your baby outward while carrying him into the garden allows him to integrate the senses of movement, the pressure (proprioception) of his limbs against your body, the visual feedback from the images of the flowers, the sounds of birds chirping all around, and the sweet smells of the blossoms. Your baby starts to create associations (and memories) between these sensations as they are activated in unison, making meaningful connections with the world around him.

Children are born with eyes that aren't yet well organized. Nature offers an array of visual stimuli for babies to observe and watch without overwhelming their visual system. As they stare at a flower or a colorful leaf, they learn how to control their eyes without crossing them. They also get to practice tracking moving objects, like a butterfly fluttering nearby or an oversized ant making its way up an anthill.

Sounds in nature can be rhythmic and calming, such as the crashing of waves, or alerting, such as the high-pitched calls of a hawk or the cry of a fisher cat. Nature sounds help children orient their bodies to the surrounding environment because they hear a variety of sounds with varying frequencies and from different distances all around them. These variations help them establish a strong sense of spatial awareness. Babies also respond to noises by turning their heads to look at the source and perhaps smiling or crying. Responding to noise is the first building block in the development of speech (Ayres 2000).

The combination of wind on the face, feelings of warmth and varying temperatures, different smells, and even tastes are all sensory stimuli that children are exposed to when outdoors on a regular basis, setting them up for healthy sensory development.

Having limited access to the great outdoors limits the amount of sensory input a child receives all at once.

Nature truly offers a complete package for your developing baby. Need more proof? Try this quick exercise: Pretend you are an infant. You are sitting in an infant carrier and watching your caregiver fold laundry in the family room. You see your caregiver moving around, you hear her singing a tune, and you smell the familiar scents of home. You even sense the occasional random movement of one or more of your limbs.

Now imagine you are being held outdoors while your caregiver is hanging up the laundry on the clothesline. She moves her body from side to side, which makes you adjust and feel pressure on different parts of your body, forcing you to use new muscles and experience new movement and position senses each time. You notice a bird fly by and follow it with your eyes. Your energetic dog Filbert comes over and picks up a chew toy in the grass, just catching your eye before he runs to the left. You hear birds tweeting in the distance, but you focus on the soft and vibrating sound coming from the caregiver's voice box close to your ears. The warmth of the sun falls on you and keeps you warm, even though a mild breeze tickles your soft skin. The smell of the laundry is close to your face, in contrast to the smell of newly cut grass. Being snuggled in your caregiver's arms brings you comfort and emotional security.

In both scenarios, the child is exposed to different stimuli. However, being outdoors offers multiple sensations at once for each sensory system. Receiving more sensations forces the sensory receptors to make more and more adaptations, leading to a more evolved and organized sensory system. These adaptations are the beginning of intelligence. All academic abilities are the end product of varied and rich sensory motor experiences during infancy and early childhood (Ayres 2000).

Time Outdoors Calms Babies

Nature is an ideal sensory environment for very young babies. Not only does it stimulate and challenge the senses to continually adapt and evolve into more complex systems, but it also does not overwhelm them. On the contrary: nature has been known to facilitate a peaceful, calm state in babies. There is good reason why your doctor recommends taking your colicky baby outdoors to induce a peaceful state. Parents and caregivers in Nordic countries (northern Europe, Sweden, Finland, Norway, Denmark) know this and have been placing babies outdoors in prams (baby carriages) to sleep for years, regardless of most temperatures and inclement weather.

Swedish day cares even put babies outdoors to sleep in the coldest of temperatures. One teacher reassured a BBC reporter that if the temperature drops below five degrees Fahrenheit, they'll put a blanket over the pram to protect the babies, but for the most part, babies remain outdoors. It is common to see rows lined up in the snow at nap time. The Swedes do this to boost the children's health, promote physical toughness, as well as reduce the children's exposure to germs and colds that spread more easily inside. Not only do caregivers feel that fresh air is good for children, but according to one research study and survey of Finnish parents, children take longer naps when outdoors (Tourula, Isola, and Hassi 2008).

The Finnish Ministry of Labour specifically recommends this practice. Its guide for parents urges them to have their children nap outdoors in prams, even in temperatures below zero degrees Fahrenheit and as early as two weeks of age. The ministry feels that babies sleep better outdoors in the fresh air than in bedrooms. The ministry also believes there is no danger for infants sleeping outdoors; they simply need to be dressed for the weather (Lee 2013). Most of the preschools and day cares in Nordic countries allow babies ample time, if not most of their time, outdoors—whether they are awake or asleep.

Many of our ideas about parenting tend to be culturally bound. It is easy to partake in a bird's-eye view of what we think is needed to foster healthy development in children by simply looking at typical parenting practices in our region. We easily get caught up in the latest magazine and newspaper articles about the top child-rearing philosophies. However, to raise children to their full potential, we'd be better off broadening our view and pulling the best practices and research from many different cultures to determine the best approach to raising strong and healthy children. Sometimes this takes having an open mind and a willingness to search for answers in other countries as well as our own.

"Floor Time" Outdoors

Spending daily and frequent time on the floor is essential for developing the muscles of the core and lower and upper body. As children push against the ground, they work against gravity, establishing postural control and developing a strong proprioceptive sense. These become the foundation for stability, effective gross motor coordination, and later development of fine motor skills. Placing babies on the ground outdoors—whether on a blanket or in the grass—enhances their sensory experience immensely.

- Young babies who spend time on their backs and bellies on the ground have opportunities to move their arms and legs freely to interact with the world around them. They can reach for blades of grass and brush their hands across it to enjoy the tingling sensation it ignites on sensitive palms. They can shift the dirt around in front of them, feeling the fine little granules on their fingertips. These sensations develop and refine the sense of touch in hands, feet, and other body parts that have contact with the natural world.

- If babies are on their bellies, turning their head to watch bugs and birds fly by further develops basic neck and eye muscle control, both of which are needed for looking and listening skills. You don't need to go out of your way to set up sensory experiences for your child; Mother Nature already has this covered. The soft wind, the swaying of grass and plants, the warmth of the sun, the novelty of hearing new nature sounds, and the moving bugs all engage the senses, inviting children to pay attention and interact with the environment. Connections are made and memories are formed. With each new interaction, children adapt and learn from the experience, furthering their motor and sensory skills each time.

Ways to Get Infants Outdoors

In summary, spending time outdoors helps young babies maintain their health and well-being, develop strong bones and muscles, calm down, and integrate as well as organize their sensory systems, laying the groundwork for emotional regulation and more complex neurological skills and motor skills. The following are ways to get your baby outdoors.

Carry them. Carry your baby outdoors on a regular basis, especially when she is awake. Save the carrying devices for when the baby is sleeping. When she's awake, carrying offers close parent-child contact, which is important for her so she can establish a strong bond and learn how to regulate emotions. Also, carrying your baby offers a greater variety of positions that she can be held in while moving, fostering a healthy sense of movement and gravity.

Tummy time outdoors. Don't be afraid to let your little one enjoy tummy time on the grass, especially once he can hold his head up for extended periods of time. Belly time on the grass offers a great sensory experience. With so many things to look at outside, he'll be motivated to push up onto his elbows to look at the world around him.

Touching new things. If babies are on their back or belly on the ground, they are likely to explore their surroundings using their hands, feet, and mouth. This is how they start to process the sensations of touch. Also, go up to trees and plants when walking with your baby. Let her feel the rough bark against her soft palms, grab hold of a leaf, and swat at a flower. If you walk by a small body of water, dangle your baby's feet in the water so she can feel the sensations of wet and cool. Let her stomp her feet and splash the water.

Napping outdoors. Make like the Nordic families and let your baby take naps outdoors. Not only will he reap the benefits of fresh air, but he will most likely sleep longer too. An American mother told me that she decided to take the advice of her Swedish aunt and put her baby outside to sleep. The baby was "out cold" and slept more than an hour longer than typical. Just be sure to dress your baby in weather-appropriate clothing.

Eating outdoors. Whether breast-feeding or bottle-feeding, there is no reason why you can't do so outdoors. Being outside allows you to go on more adventures with baby in tow without having to find an indoor area to feed. Also, babies enjoy the peacefulness that being in nature offers, soothing them while they drink.

Increasing overall time outside. The more time you spend outdoors with your baby, the better. Let her enjoy the fresh air, the sights around her, crushing dried and noisy leaves in her hands, feeling the wind softly blow against her skin, and the cold air of frosty mornings nipping against her nose. Old and new sensations mix together to create new experiences, new memories, and new challenges to overcome. The opportunities for sensory and neurological growth are endless in the outdoors. Nature offers the ultimate sensory experience for babies to explore and gain new skills from.

BABIES IN NATURE (SEVEN TO TWELVE MONTHS)

Older babies experience the excitement of moving in new and different ways. Varied movement experiences are especially important between seven and twelve months of age to help babies master body awareness and conquer new sensory motor skills. They start to enjoy more rigorous movement experiences, such as being thrown into the air, getting horsey rides on a caregiver's back, being spun around in circles, and being danced around in a caregiver's arms. Creeping and crawling on hands and knees integrate many senses and allow babies to feel independent enough to explore their surroundings on their own (Ayres 2000).

Moving around in nature offers many new challenges to the mobile baby. Instead of the continuous smooth surface that most indoor environments offer, crawling and creeping outdoors offers a rich and varied sensory experience, further enhancing sensory organization.

Play Outdoors Develops Competence

Play outdoors is essential for a growing baby. It is through large, full-body movements that children start to develop motor and sensory maps, leading to effective spatial awareness. Through the manipulation of small things outdoors, such as acorns and blades of grass, babies learn how to effectively use their hands and fingers. By crawling over the uneven terrain typical of the outdoors, infants start to develop balance, stability, and coordination. Dr. A. Jean Ayres writes, "Play expands competence. The child may not need this competence until later in life, but he won't develop much competence unless he plays effectively as a child" (2000).

Babies, just like older children, need time to move about outdoors on a daily basis. They need to challenge their bodies, minds, and senses by exploring the world around them.

UNEVEN TERRAIN CHALLENGES GROWING INFANTS

The terrain outdoors is not level or smooth. Its properties are constantly changing. It goes from hard to soft to somewhere in between. It changes from warm to cold to moist to dry. It goes up and dips back down. These changes challenge a baby's sensory system to adjust and adapt while he or she is crawling and learning to walk. The more adaptations required, the more organized and refined the senses become. At the same time, the varying tactile and temperature sensations increase the baby's tolerance to these experiences as well.

Let's do another exercise. Pretend you are eleven months old. Imagine crawling outdoors. You start off on a relatively flat dirt path. Your older brother is playing nearby, and he catches your eye. You start to crawl in his direction. You have pants on but can feel the hardness of the ground through them. The pebbles and compact dirt push up into the palms of your hands as you move.

This offers great tactile feedback, and you begin to discriminate between the relatively big pebbles and the smaller granules of dirt. You finally make it to the grass and instantly feel a relief of pressure on your hands and knees as you start to navigate the much lumpier and cushier terrain. Then you start to go down a slight incline, which really forces your muscles and position senses to adapt.

The hill is steeper than you are used to, and you momentarily lose your balance, falling onto your chest and belly. Your shirt slides up and you feel the soft tickle of grass on your stomach. Determined to get to your brother, who is now climbing a tree, you get back on all fours and continue your journey on this uneven terrain. The changing pressure on your hands expertly molds and refines the arches in your palms, which are critical for developing more precise fine motor skills.

The warm sun beats down on your back as you make your journey. However, the breeze offers relief, and you eventually make it to a more shaded area where a few trees loom overhead. You look up at one of them and this time fall onto your side. Constantly falling, adjusting the body, and getting back up again helps you to develop good spatial awareness, further enhancing your ability to maneuver the environment with precision. You finally make it to where your brother is.

The ground is soft and full of gentle pine needles. You crawl up to the big oak tree your brother sits in, high off the ground. You shift the weight to your hips and legs, which allows you to reach up and hold on to the hard, rough, grooved bark running down the tree. Both palms grab hold of the rugged tree, and you hoist yourself into a standing position. You look up at your brother while struggling to maintain your balance. He looks down at you and says your name. The smell of the bark lingers as you smile back at your brother. A connection is made. A memory is formed.

The more practice babies have moving over uneven terrain while experiencing changing environmental sensations, the more refined and capable they will be in how they move and regulate sensations. Variation in the gravitational forces and loads on different parts of the body creates stronger bones and muscles. It also increases the amount of input the joints and muscles get, improving their ability to regulate the distance and force needed to execute movements effectively. Changing direction also sends maximum stimulation to the inner ear, helping develop a skilled balance system. The more babies move, the more they fall, the more sensations they experience, and the more they are able to master new motor skills and further integrate and organize senses. The result is a strong and skilled child.

LAYING THE FOUNDATION FOR LANGUAGE AND MEMORIES

Play outdoors provides meaningful experiences, laying the foundation for speech development and memory formation. The combination of rich, new sensations, such as sights, movements, and smells, helps create strong associations between an experience and the meaning of that experience. Since nature offers a multisensory experience, it creates more feedback about an object or situation, leading to a more advanced processing of that particular object or situation in the brain.

For example, when a baby reaches to touch and then process the soft fur on an alpaca, he is also processing the strong smell of the farm, the alpaca's response as it leans toward him, and finally the soft moans the alpaca makes. All these sensations help the child fully process the experience and support basic speech as the child tries to mimic the alpaca's soft groans. A connection is formed and the child will likely remember this experience. The next time he sees an alpaca, he will almost certainly make soft moaning noises at it.

Ways to Get Older Babies and Toddlers Outdoors

Nature is as essential to our sensory systems as good nutrition is to our health. We need to spend time in nature to make sense of the world around us and to grow a more advanced and capable sensory motor system—the foundation for all academic skills. Therefore, it is critical that we allow our babies to spend time outdoors at an early age. Below are some ideas and suggestions for getting older babies outdoors.

Continue to carry them. Allow your baby to bear as much of her own weight as she can when you walk outdoors with her. As she turns to look at something, she will shift her body; allow her this freedom of movement in your arms. The pressure of her body against yours is an organizing sensation. It provides important input to muscles and joints, ultimately improving body awareness.

Also, holding on to caregivers allows babies to strengthen the muscles of the shoulder girdle and core (stomach, back, and diaphragm). These muscles need to be developed for proper stabilization, which supports higher-level sensory organization, such as regulating emotions and arousal levels, coordinating movements in a smooth and efficient manner, and having good spatial awareness.

Let them explore the yard. Babies don't need a lot of space to explore. Just a patch of grass will suffice. Allow them to climb onto rocks, roll in the grass and pull up blades of it, and crawl under low branches or into nooks and crannies outside. These experiences will help them practice motor skills while learning about their body and the space around them.

Play in the mud. Sensory bins are all the rage. A "sensory bin" is a simple container filled with a tactile substance, such as rice, beans, sand, or water. Often a child sits or stands in front of the sensory bin and plays with the substance using his hands. Sensory bins are a way to bring sensory experiences for babies inside. However, they are time-consuming, cost money, have limited space, and often create a mess. Letting your baby play in a mud puddle is free and natural, and there is no clean up—except for your baby! Playing in the mud allows babies to fully immerse their body in a meaningful and engaging experience. When you add the elements of sun, light rain, wind, and nature sounds, the sensory experience is enhanced in ways that no sensory bin could ever accomplish.

To encourage more play, set up pots and pans with wooden spoons near the muddy area. Babies will experiment with the mud by throwing it, placing the sticky mess in containers, emptying the containers, kicking and splashing in the puddle, and squealing with delight. Playing in a mud puddle is sure to be a fun and playful experience.

Play in the rain. Along the same lines, let your baby play in the rain. Make sure the weather is warm enough so your baby doesn't get too cold. Let her feel the warm and gentle rain fall lightly upon her arms and forehead. Set her down *beside* rain puddles (not inside the puddles). Allow her to decide whether or not she wants to go in the puddle. Offer things that float for your baby to watch and pick up. This is a whole-body experience that allows your baby to explore the sensations of wet in a meaningful way.

Explore the beach. The beach offers babies an array of sensations. Allow them to crawl around and explore the feeling of sand on their hands and knees as they navigate the uneven but soft terrain. The smells of the ocean, the soothing crashing of waves, as well as the calls of seabirds are all sensations that put babies in an alert but calm state as they explore their surroundings. Try giving your baby containers of different sizes that he can fill with water and sand.

Picnic in a park. If you don't have lots of nature around your home, go to a park for the day. You may want to bring a lunch, snacks, and water so you can extend your stay even longer. State and national parks are especially fun places to bring little ones to. They are often well maintained and clean. Set your child down in the grass and let her roam and explore. Also, bringing children to parks can expose them to new sights and sounds that they wouldn't see or hear around the house, such as those of waterfalls, woods, and rock formations.

A LITTLE LESS NO, A LITTLE MORE YES

In chapter five we learned that by constantly telling young children no we prevent them from gaining rich sensory experiences and attaining the necessary building blocks to foster healthy sensory development. The same goes for babies. They need daily opportunities to challenge their bodies and take risks, such as picking up sticks, crawling on uneven ground, and jumping off the bottom step of a flight of stairs, in order to process the functions of their body and make sense of the world around them.

Dr. A. Jean Ayres calls this their "inner drive." She believes within every child there is great inner drive to develop sensory integration. If left to their own devices, babies seek out the sensory input they need to master a certain skill (Ayres 2000). For instance, older babies may practice crawling onto and off of a low rock, over and over, until they can do so quickly and efficiently. Once they conquer this skill, they may try the next level in development, such as sitting or standing on top of the rock or crawling backward onto the rock.

Our job as parents is to support our baby's growing independence and increased need to move. Be there just in case your baby needs you, but encourage him to try new things by simply being present. There's no need to say anything. Your presence and a smile will reassure your baby enough to take new risks.

At the same time, it is important to never place babies on high equipment if they aren't able to get there themselves. They may not have yet developed the strength and adequate body awareness to effectively manage an elevated environment. Placing them in such a situation too early could lead to a fall and a sense of failure. Therefore, follow your child's lead; she will show you what she is ready to do and what she is not yet capable of doing.

By now you are probably starting to realize the many, many benefits of letting babies play and explore outdoors. However, some of you may still be reluctant to let your baby outdoors due to a number of preconceived worries or fears. Things like poisonous plants, biting bugs, and potential injuries are enough to give anyone pause. However, the "dangers" of nature may actually pale in comparison to the many dangers present in the household. Everything from toxic substances in your cabinets to sharp knives just out of reach are hazardous to babies as well. Knowing our homes pose such risks doesn't keep us away from them. On the contrary, we simply implement safety measures, such as putting child-safety locks on the cabinets and keeping sharp objects out of reach.

We need to do the same thing with the great outdoors. Instead of running away from nature, babies would be better served if we educated ourselves about what is and isn't harmful and took appropriate precautions. I covered a substantial list of common sense safety tips in chapter five; however, let's touch upon three of the most common ones for babies.

- **Boo-boos and splinters.** The idea of a bruise or a splinter is very abstract to ten-month-olds. But an actual scrape or prickle on a fingertip is *real* and leads them to a more complete understanding of their surroundings. Babies learn valuable lessons from these, such as cause and effect. *If I touch that prickly bush, it will pinch me.* The next time they encounter the same plant, they will be less likely to touch it. Boo-boos and splinters also teach them how to regulate emotions, such as fear and frustration, as well as how to tolerate sensations of pain.

- **Getting dirty.** When you see children covered in dirt with leaves in their hair, don't scold them for getting dirty. As we learned in chapters two and four, playing with dirt and getting messy helps them to refine and develop a strong tactile sense. Chapter four touches upon the hygiene hypothesis and how sampling a little dirt and playing with it actually improves the immune system, providing invaluable protection against the development of allergies and asthma.

- **Eating dirt and chewing on sticks.** The saying "a little dirt never killed anyone" comes to mind when I think about children consuming natural products of the outdoors. Ingesting dirt, sand, or grass will not harm children. In fact, it is less likely to harm them than putting lotion, cleaning solutions, or other household items in their mouth. Babies are very oral. They use the oral sense to learn more about an object's size, texture, temperature, and taste. Putting pinecones, a little dirt, or a stick in the mouth teaches them about the natural world, and the exposure to germs improves their immune system.

Encourage your baby to explore the world through all of the senses. However, if you have a crawling baby, make sure to keep an eye out for trash, animal droppings, bugs, plants, and small objects like rocks that could get lodged in the throat; keep babies away from these items.

IN A NUTSHELL

Above all else, enjoy your baby! The infant phase is very short. Contrary to societal norms, the first year of life is one of the most critical times to get your child outdoors. During their first few years of life, children make rapid neurological connections and associations between their senses, creating meaning that helps them understand the world around them. Don't wait until they are older to let them explore and take risks. Go outdoors with your littlest ones and let them explore their surroundings. With the right amount of time and experience in nature, children's sensory and motor systems will thrive, laying a strong foundation for later life challenges and academic learning.

CHAPTER 9

Getting Children to Play Creatively and Independently Outdoors

My girls, ages seven and nine at the time, were getting on each other's nerves—*again*. "Stop it!" One griped. "No, you stop it!" The other yelled back. They were sitting on the swings, and the youngest was defiantly humming a tune over and over again. My other daughter, clearly agitated, plugged her ears and groaned. They bickered back and forth for a while. Finally, my eldest came up to me and asked, "Can I have a playdate?"

Knowing that a playdate was out of the question due to time constraints, I told her, "Why don't you two ride down to the Johnsons' home to see if the boys want to play?" The Johnsons lived about a five-minute bike ride from our house. The girls would have to ride down a dirt road in order to get there. "Yeah!" She shouted back. "Can we?" Since it was their first time biking down there, I texted my friend to let her know the girls would be coming, and that she should send them home if her boys couldn't play.

It was my friend's first time letting her boys have this kind of autonomy, too. By letting our kids enjoy this simple act of freedom, a whole new world opened up for them. The offering of autonomy was liberating, and we were both excited for their newfound independence. After that, my daughters started biking to other friends' houses. More and more often we had children coming to our house on bikes and then taking off again to go to the rock pit, to build dams in the stream, or to play elaborate games outdoors. They ventured farther and got more creative with their play schemes. I simply had to let go of my stranger danger and overcome my fear of the kids getting hurt in order to let it happen. No longer did I have to plan every outing, every playdate.

Overcoming parental fears may not be your biggest issue when it comes to allowing your child to have independence. Perhaps it is your child's boredom and lack of motivation getting in the way. I hear from many parents that even if they let their children play outdoors, their kids often have a hard time knowing what to do. When children aren't given enough opportunities to practice play outdoors, they often lack the confidence and skills needed to become independent and creative with their play. They quickly become bored and want to go back indoors—into their comfort zone.

This chapter is designed to help you overcome barriers (boredom and parental fear) to fostering creativity and independence in your child in an outdoor setting and provide you with ways to do so, whether you live in the suburbs, a rural area, or an urban environment. The key is letting go of fear, providing ample opportunity for creative play outdoors, and taking baby steps toward allowing your child more independence.

OVERCOMING BARRIERS TO INDEPENDENT PLAY

Overcoming your fears and letting go of the notion that you are responsible for preventing boredom are the first steps toward fostering independence in your child. In fact, when children experience boredom, it often forces them to get creative with imaginative play.

Overcoming Fear

A recent poll taken in the United Kingdom found that 53 percent of parents surveyed blame traffic for their reluctance to let children play outdoors independently. 40 percent fear that strangers will snatch their children. Still others worry about what the neighbors will think or are afraid their children may get hurt. Most parents admitted that if they saw more children playing outdoors, they would be more apt to let their own children outdoors (Levy 2013). What are your fears? What is getting in the way of you letting your children have the freedom to, on their own, make a new friend who lives down the street? Of being able to play a game of pickup basketball with the kids who live in your area?

We learned in chapter five that, contrary to popular belief, the world is just as safe, if not safer, than it was thirty or more years ago. Here are a few ways to ease any lingering fears you may have.

- **Remember that independence now equals safer children in the long run.** Remind yourself that promoting independence now makes for more capable, knowledgeable, and self-sufficient children in the long term.

- **Start simple.** Don't do anything you are not comfortable with. Start with small steps by encouraging independence. Things like staying outside while the kids play and allowing them to roam a little farther each time are strategies I discuss in more detail later in this chapter.

- **Observe and practice.** Have your child prove that he can handle walking or biking to a neighbor's house by going with him the first few times. Observe your child playing outdoors from a distance to see how he is doing without him knowing it. Sometimes just seeing how your child handles different situations on his own can help you be more comfortable letting him have the space he needs to develop independence.

- **Set clear guidelines.** Be specific on when you want your child home; how far away she is allowed to walk, bike, or play; and whether or not you want her to check in with you, and how often. I discuss these things in more detail later in this chapter.

Moving Beyond Boredom

Do you often worry that if your children get bored they might get into mischief? Or that your children will annoy each other silly and in turn drive you crazy? You are not the only one. Many parents today work extremely hard to make sure there isn't idle time after school, on the weekends, and during the summer. We set up playdates for our children, sign them up for sports, line up vacations, and put them in camps—all to prevent downtime. However, by preventing opportunities for boredom, we are avoiding the very thing they need to experience in order to become independent and creative with their outdoor play.

Children need to get bored from time to time in order to learn how to move beyond these feelings and come up with their own play ideas. When children get bored, we often feel that it is *our* job to entertain them and offer an assortment of activities. Instead, remember these simple tips.

- **Children need to experience boredom.** Oftentimes, it takes a period of boredom before children start seeing objects in new ways, such as using a stick for a mixing spoon or pretending that a puddle is hot lava. Boredom forces kids to come up with more complex levels of play.

- **Let children daydream.** When children are allowed quiet time to ponder life and what they want to play, they start to generate some of their most creative ideas.

- **Let go of preconceived notions.** If your children are walking around aimlessly, let them. If they are seated and quietly looking at something, let them. Sometimes what we have in mind as "play" looks different than what children consider play. Also, children need time to explore and think before they get into more creative types of play.

- **Open up your calendar.** Allow children opportunities to practice overcoming boredom by opening up your schedule. Resist the urge to have to do something every day during vacations and weekends. For every "free" day that you have back-to-back activities scheduled, try having a day off during which your family has nothing planned. For instance, if you are going to the beach on Saturday, make sure you stay home and relax for most of Sunday.

WAYS TO ENCOURAGE INDEPENDENT AND CREATIVE PLAY

We all want our children to be independent at some point. However, our expectations about when this should occur vary widely. When some parents think of independence, they automatically think of the teenage years: driving a car, working a first job, and selecting a career path. However, to wait until the teen years before encouraging independence would be a great disservice to our kids. They really need to learn independence at an earlier age. They need to feel a sense of choice and stretch those decision-making skills even as infants. In fact, if they aren't given a sense of freedom at an early age, children can develop resentment, anxiety, heightened sensory aversions, picky eating habits, and behavior issues as there becomes a need and battle for control. Children need to feel safe by having natural limits set for them, but they also need to have space and time to play, unhindered by adult fears, in order to truly flourish. The space and time to play are fundamental to encouraging independent play outdoors.

Giving Children Space

Giving children space may be one of the hardest things for parents to do. We love our children so much. We just want what is best for them. We'd like to be there every moment of their life—to capture every milestone and every new skill in a snapshot in our minds. We don't want to miss out. However, just like we need our own space away from work, busy schedules, and people in general, our children need their own space too.

Children benefit from time alone to reflect on life, to play, and to test out different theories without a parent hovering over their shoulder observing their every move. They need a break from being watched. In order to provide that space, loosen the reins a little and step back. Allow freedom to happen. Although stepping back may be difficult to do at first, with baby steps it can progress naturally and become comfortable for both you and your child.

It's critical to allow children opportunities to play on their own and to formulate their own play schemes. If we always feel obligated to play with our children, they may become dependent on our guidance and support in order to play imaginatively. How do you offer your child space? First, determine what age your child is ready for space. Second, determine what abilities your child must possess in order to be offered space. And third, determine what space is appropriate and available. Depending on where you live, these steps may take a little more creativity. Below are examples of "spaces" to consider.

THE BACKYARD

Go outside into your backyard with your child and just be present. Your child will feel comforted that you are within viewing distance while he explores his surroundings and plays. This is a great time to get outside chores done. Here are a few examples of ways to foster independence.

- **Start a garden.** As you garden, your child may be inspired to dig in the dirt, explore it with a shovel, and even find worms. He may eventually wander off to investigate a hole in a nearby tree or look under rocks for insects, demonstrating the beginning of independent play.

- **Trim bushes.** As you trim the bushes and create piles of branches, your child may be inspired to use the branches to create a fort or bridge or use them in imaginary play (for example, a branch may become a walking stick for an old wizard or a tool with which to dig for dinosaur bones).

- **Rake leaves.** As you rake leaves into bins of yard waste, be sure to leave piles for your child to explore. Let him bury himself, throw the leaves up in the air, or jump into the piles, practicing independence while you rake.

- **Play games from time to time.** When letting your child explore the backyard independently, especially in the beginning, feel free to join him in play, such as wrestling, swinging, or playing catch. Spending time with your child shows your playful side as well as helps you bond with him at a deeper level.

- **Build something together.** Have your child help research the play equipment he'd like in the yard, and then build it together. If your child helps plan and build the structure, he's more likely to use it.

- **Bring a smaller project outdoors.** Work on your latest knitting or scrapbook project outdoors. Doing so allows you to be physically present outdoors while your child plays in the yard.

THE PARK

For children growing up in urban areas, the park may be just the place they need to practice playing independently at a young age. At first you'll want to stay close by while your child plays. Simply being present often gives your child the confidence and reassurance she needs in order to start exploring her surroundings on her own. Here are a few things to keep in mind when fostering independence in urban areas.

- **Seek out natural areas.** Try to find parks that offer enriching natural elements to explore, such as areas with water, dirt, grass, and woods.

- **Bring a good book to read.** While your child plays nearby in the dirt or amongst the trees, read or knit. Try to take some time for yourself, and gradually refrain from watching her every move.

- **Bring other children.** Oftentimes, having other children with you can inspire your own child to play more independently in a new situation or place.

- **Stay close when near water.** When you are near a lake or a stream, stay close by to supervise young children. As they get older, you can distance yourself. However, it is still important to keep children within sight any time they are near water.

THE STREETS

When your child is older (ages nine and up) and mature enough to know basic directions, he may be ready to walk or ride his bike, scooter, or skateboard to the neighbor's house down the street or to a nearby store with friends. Put some guidelines in place and teach him basic street smarts before letting him go anywhere. Here are a few tips to help keep your child safe and, therefore, increase his ability to go places independently.

- **Teach basic navigation skills.** Educate your child on how to get around his neighborhood or navigate city streets without getting lost.

- **Clarify what traffic signs mean.** Teach your child what different traffic signs signify and what he should do when he sees them.

- **Practice first.** Practice crossing and navigating the streets with your child over and over until he knows exactly what to do, demonstrates competence, and exhibits good safety awareness.

- **Determine boundaries clearly.** Give your child clear boundaries before letting him go off on his own. For example, you may tell him, "Don't go past Tony's house or the library."

- **Consider a walkie-talkie.** With a walkie-talkie you can communicate with your child as he heads somewhere, when he gets to his destination, or as he's coming home.

- **Eat with your kids.** Have your child come home at lunchtime to check in with you and to eat lunch.

- **Have your child check in.** Your child can call when he gets to his friend's house or check in every few hours.

- **Travel with other kids.** Making sure your child has one or two friends or siblings with him is always a good idea in case one child gets hurt or needs help.

- **Pack snacks and water.** Pack a small backpack with snacks and water that your child can comfortably carry, or equip his bike with a basket to hold these refreshments. By doing so, your child will stay fueled up and hydrated.

- **Teach your child "street smarts."** For example:

  - What places to avoid and why.

  - When to talk to strangers.

  - If help from a stranger is needed, who is the best person to look for (for example, a police officer).

  - If your child gets lost, what he should do.

  - If your child is in trouble, what he should do.

  - If your child gets hurt, what he should do.

  - To never get in a car with a stranger or go anywhere with a stranger.

When introducing independence to your child, start small in the beginning. Have your child bike just to the next-door neighbor's house. Increase the boundaries from there. Every child is different. Some children may not be ready for this sort of independence. Until they are, teach them street smarts and accompany them on their journeys. Have them practice being independent until you feel confident they can navigate the streets with success. By implementing the simple strategies in the lists above, children can and should learn how to navigate their neighborhood or more urban streets with ease and confidence.

THE WOODS

If you are lucky enough to have a patch of woods nearby, it offers a great place for children to explore and play. As we learned in chapter four, playing amongst trees tends to have a calming effect on children. It also inspires creative play. A wooded area provides an enchanted atmosphere that ignites the imagination and gets children away from the adult world. Here are a few tips to help keep children safe and, therefore, increase their ability to go places independently.

- **Go with them at first.** As with the backyard, park, and more urban spaces, you can introduce play in the woods by remaining close by at first. Bring a good book to read while your child plays in the woods.

- **Pick up sticks.** Try picking up sticks on the forest floor while your child plays. Doing so will keep you busy while she explores. Also, by cleaning up the forest floor, especially if the woods are yours, you open up the sense of space in the woods, which can prompt children to venture even farther.

- **Allow for roaming.** As children become more and more independent and mature with age and competence, gradually give them more space. Consider allowing them to go past your line of vision while still listening for their voices, and check in on them once in a while without them knowing.

- **Explore with friends.** Eventually, if you live close to the woods and your child is fully competent at knowing how to get in and out of the woods, you can let her go with friends or siblings.

- **Establish physical boundaries.** It is important to establish clear boundaries so your child doesn't get lost in the woods. Consider adding bright markers to create paths or to outline a physical boundary.

- **Point out hazardous plants and animals.** As we learned in chapter five, it's a good idea to teach children what potentially hazardous plants, animals, and insects to avoid.

- **Give clear expectations.** Have your child come home for lunch or carry a walkie-talkie. The key is to figure out what your child is truly capable of and what measures you should put in place, in the beginning, to make sure you are comfortable with her playing independently.

Giving Children Time

Another way to foster independent play is to give children plenty of time to do so outdoors. Ideally children should have at least two to three hours of uninterrupted play outside every day. When we allow children adequate time to play without adult interference at TimberNook, we observe significant changes in the way children play over a short period of time, sometimes in just a week. The first few days of camp, children play in simplistic ways. They usually walk around and explore what is in their environment. Often, they line up in front of the rope swing or attempt to climb trees. They look for frogs and try the ropes course. They often come up and ask us what is next on the "schedule" and check in to make sure they can do different activities, such as playing with sticks or climbing on top of boulders.

Not until the middle of the week—or even the end of the week for some kids—do they start to engage in more creative, independent play. This is especially true for kids who are new to TimberNook. Children who have been coming for years tend to engage in independent play much more quickly, and their play develops more complexity early in the week.

We also notice that it consistently takes children about forty-five minutes the first few days of camp to get into "deep" play. Deep play is a type of play that involves purpose and is not merely exploratory in nature. By the time children have engaged in deep play, they have chosen who they are going to play with, have chosen what they are going to play, have taken on their roles, and have developed their play arrangements. We can almost hear when this happens at TimberNook, because the woods become quiet and peaceful. Children are dispersed throughout the woods engrossed in their own play schemes.

As the week goes on, children dive into deep play at a much faster rate. By the end of the week, they are in deep play within minutes, jumping right into their roles and games with peers. However, children need time, patience, and the opportunity to step back from adults for this to occur. Playing at a deeper level takes the right environment and ample opportunities to practice the skill.

Are you worried that it may be difficult to add at least forty-five minutes of uninterrupted outdoor play each day? Here are some tips for carving out more time on a daily basis.

- **Play after school.** Allow your child free play immediately after school. If lessons or sports practice are necessary, try to schedule them later in the day or on the weekends.

- **Play in the morning.** Many parents allow their children to watch cartoons in the morning before school. Consider allowing them to play outdoors instead.

- **Make time on the weekends.** Allow for one day on the weekend that the family stays home. While the children play outside is a good opportunity to get caught up on chores.

- **Rethink recess.** Work with school administrators and teachers to see if longer recesses are a possibility. Or inquire about the possibility of adding recess before school. Refer to chapter seven to learn about the importance of recess for child development and strategies on how to make recess an optimal play experience.

Bringing Friends into the Picture

When children are around other children, they naturally inspire each other. I see this all the time with my own children. When it is just my two daughters at home, they tend to play mostly in the yard. Sometimes they play in mud puddles, but mostly they play on the swing and slide and ride bikes. When they have friends over, they roam farther, play in the woods, build dams, and get more creative with their play. Not sure how to get more children in the picture? Here are some simple tips.

- **Schedule all-day playdates.** Having children come over for the full day allows for plenty of time to get bored, eat meals together, explore, imagine, and create new play opportunities.

- **Allow your children to roam.** By allowing your child to bike to friends' houses, other families will often be inspired to let their children also ride bikes and roam a little farther.

- **Find like-minded parents.** Seek out parents who also let their children play outdoors and encourage independence. These children are more likely to play outdoors with ease and confidence, which can inspire the same in your kids.

- **Create an "open" house policy.** Pick a day once a month or once a week during which other children are allowed to come and go without having to call ahead or make arrangements. Or establish an open-house policy all the time, depending on what works for your family.

Using the Environment as Inspiration

Another way to inspire creative and independent play is by providing a rich and inspiring environment. Environments that offer varied elements, such as a stream to explore, muddy areas to get messy in, or woods to play in, enrich and stimulate children's play in different ways. Here are just a few examples of how play can vary depending on the environment.

- **Play near a stream.** Streams can encourage children to build a dam, make a bridge to walk across, or create "boats" to float down the stream.

- **Play at the beach.** At the beach, children often test their hand at building sand castles or sculptures, playing at the water's edge, or experiencing risk taking by swimming and diving into to the waves.

- **Play in the woods.** While playing in a wooded area, children may pick up large sticks to create a fort or little ones to build a fairy house.

- **Play near mud puddles.** While playing near a mud puddle, children may seek out frogs to catch, create mud sculptures, try going barefoot, build a bridge, float items in the water, or throw objects into the puddle to make a big splash. They may even try mud slides—sliding into the puddles on their bellies.

LOOSE PARTS

As we discussed in chapter seven, to further enhance creativity you can provide loose parts for play in the unique settings discussed above or even your own backyard.

- **Avoid plastic or childlike objects.** These toys typically promote one type of play, and the goal of free play is to get children using their imagination with items they might not ordinarily play with.

- **Offer building items.** Planks of wood, tires, cut pieces of tubing, tree cookies, pieces of fabric (such as translucent sheets), and sticks often provide the foundation for imaginary play. For instance, children may use these items to create a house and then play house with smaller loose parts, such as pots, pans, shells, and baskets.

- **Visit garage sales.** Yard sales and thrift shops offer a treasure trove of interesting objects for little money. Look for stainless steel scoopers, pie pans, serving pitchers, metal colanders, metal trays, baskets, costume jewelry, and so on. Allowing access to items primarily used by adults often excites children and encourages them to play in new ways. These items can be used over and over before needing to be replaced.

- **Change up the loose parts.** Rotating the objects your children play with gives them new ideas and builds the repertoire of things they can play.

- **Keep quiet.** When placing out loose parts, it is best not to say anything about them. Let the children decide how they want to use the items. We don't want to limit their play by offering our own ideas; children will become dependent on our ideas and thoughts. The whole point of loose parts is to inspire independent and creative thinking. They may come up with something we would never have thought of. Simply find a place to watch them play from a distance.

Providing Simplicity

It is important to keep the play environment simple. Sometimes too much equipment or even too many loose parts can distract or overwhelm children, deterring them from playing creatively and deeply. Here are a few tips to keep things simple while fostering creative play in your backyard.

- **Get rid of the giant play structure in the backyard.** Large play structures often become the center of attention and discourage children from playing in the rest of the yard.

- **Offer a few pieces of equipment.** Spread a few pieces of equipment in the yard to encourage exploration of these areas.

- **Offer stimulating and simple equipment.** Place a long stainless steel slide down a natural slope. In another area of the yard, put up a rope or tire swing with a long span that kids of all ages can use.

- **Install a sand pit.** Sand and dirt are always good to have around to encourage healthy sensory (touch and proprioceptive) play. Having a well pump nearby is another fun idea to foster creative mud and water exploration. Children not only love pumping water into containers of different sizes, but when a water source is placed by sand, they often mix the two substances to make creative concoctions like mud soup and mud cakes.

- **Loose parts.** Loose parts can also be used and rotated in backyards to inspire creative play. As we learned in chapter seven, keep the loose parts for basic building (planks of wood, tires, bricks) outdoors all the time. However, only offer a few of the smaller loose parts (baskets, shells, masks) at a time.

IN A NUTSHELL

Overcoming our fears and the tendency to keep children entertained are the first steps toward encouraging active, independent play outdoors. This type of play is further enhanced when children are given plenty of time for free play on a daily basis and are offered inspiring outdoor environments to explore.

To integrate independent free play in your child's life, start by offering loose parts, engaging environments, and an adult presence that does not interfere with or distract from independent play. Slowly phase out yourself and the loose parts as your child becomes more and more independent and creative. After hours of practice, the need for loose parts and an adult presence will lessen. Soon your child will be able to come up with his or her own ideas and play schemes without prompts or stimulating environments. The goal is to get your child to complete independence, which is the ability to think creatively and openly. These abilities will promote the development of a strong and capable individual for years to come.

Acknowledgments

This book marks a new chapter in my life. I would like to thank my agent, Stefanie Von Borstel, for her wise counsel and for seeing the potential in this book before anyone else. Also, Marisa Solís, my developmental editor, for her unique perspective and extraordinary suggestions that ultimately resulted in a better book. New Harbinger's Melissa Kirk for her timely support, as well as Julie Bennett, Jesse Burson, Bridget Kinsella, Nicola Skidmore, James Lainsbury, and the rest of the New Harbinger team. Hilaree Robinson, Jeremy Robinson, Jessica Carloni, and Dee Dee Debartlo also gave invaluable editorial support and advice throughout the book process.

Thank you to Dr. Nichia Faria, who is not only a friend but talked extensively with me about current pediatric health and wellness issues. Thank you to all the parents for their stories and the elementary and middle school teachers I interviewed and observed; you gave life to this book and are the purpose behind it. I am grateful for Elyse Bachrach, Steve Renner, Molly Wilson, Millissa Gass, Megan Thunburg, Megan Sharples, Scot Villeneuve, Tim Cook, and Dave Underhill for their unfailing support and help with creating TimberNook, which carries out the mission of this book. Also, my deepest thank you to Richard Louv for his inspiration and generous support of my work.

I especially thank my husband, Paul Hanscom, who believed in this book and TimberNook from the beginning. Thank you for your constant love, support, and guidance. Without you, there would be no book, no TimberNook. I'm so grateful for my loving parents, who taught me at a young age to reach for the stars. They have loved and supported me with every endeavor I've taken on. Much gratitude goes to my mother-in-law, Sue Hanscom, whom I view more as a friend than anything. Thank you for your constant encouragement and the reminders to keep things simple. And finally, to my girls, who served both as my original inspiration and my biggest cheerleaders. You fill my life with joy every day.

Recommended Reading

Biel, L., and N. K. Peske. 2009. *Raising a Sensory Smart Child: The Definitive Handbook for Helping Your Child with Sensory Processing Issues*. New York: Penguin Books.

Bronson, P., and A. Merryman. 2011. *NurtureShock: New Thinking About Children*. Boston, MA: Hachette.

Gray, P. 2013. *Free to Learn: Why Unleashing the Instinct to Play Will Make Our Children Happier, More Self-Reliant, and Better Students for Life*. New York: Basic Books.

Hirsh-Pasek, K., and R. M. Golinkoff. 2003. *Einstein Never Used Flashcards: How Our Children Really Learn—And Why They Need to Play More and Memorize Less*. Emmaus, PA: Rodale.

Jensen, E. 1998. *Teaching with the Brain in Mind*. Alexandria, VA: Association for Supervision and Curriculum Development.

Kranowitz, C. 2006. *The Out-of-Sync Child: Recognizing and Coping with Sensory Processing Disorder*. New York: Perigee Books.

Lahey, J. 2015. *The Gift of Failure: How the Best Parents Learn to Let Go So Their Children Can Succeed.* New York: HarperCollins.

Lansbury, J. 2014. *Elevating Child Care: A Guide to Respectful Parenting.* CreateSpace Independent Publishing Platform.

Louv, R. 2008. *Last Child in the Woods: Saving Our Children from Nature-Deficit Disorder.* Chapel Hill, NC: Algonquin Books.

Mogel, W. 2008. *The Blessing of a Skinned Knee: Using Jewish Teachings to Raise Self-Reliant Children.* New York: Scribner.

Payne, K., L. Llosa, and S. Lancaster. 2013. *Beyond Winning: Smart Parenting in a Toxic Sports Environment.* Guilford, CT: Lyons Press.

Payne, K., and L. Ross. 2010. *Simplicity Parenting: Using the Extraordinary Power of Less to Raise Calmer, Happier, and More Secure Kids.* New York: Ballantine Books.

Siegel, D., and T. Bryson. 2012. *The Whole-Brain Child: 12 Revolutionary Strategies to Nurture Your Child's Mind.* New York: Bantam Books.

Skenazy, L. 2009. *Free-Range Kids: How to Raise Safe, Self-Reliant Children (Without Going Nuts with Worry).* San Francisco, CA: Jossey-Bass.

Tough, P. 2012. *How Children Succeed: Grit, Curiosity, and the Hidden Power of Character.* Boston, MA: Mariner Books.

Ward, J. 2008. *I Love Dirt!: 52 Activities to Help You and Your Kids Discover the Wonders of Nature.* Boulder, CO: Roost Books.

References

Alter, A. 2013. "How Nature Resets our Minds and Bodies." *Atlantic*, March. http://www.theatlantic.com/health/archive/2013/03/how-nature-resets-our-minds-and-bodies/274455.

Alvarsson, J. J., S. Wiens, and M. E. Nilsson. 2010. "Stress Recovery During Exposure to Nature Sound and Environmental Noise." *International Journal of Environmental Research and Public Health* 7 (3): 1036–1046.

American Academy of Pediatrics. 2013. "Managing Media: We Need a Plan." October 28. https://www.aap.org/en-us/about-the-aap/aap-press-room/pages/Managing-Media-We-Need-a-Plan.aspx.

American Physical Therapy Association. 2008. "Lack of 'Tummy Time' Leads to Motor Delays in Infants, PTs Say." News release, August 6. Retrieved from http://www.apta.org/Media/Releases/Consumer/2008/8/6.

Asher, M. I., S. Montefort, B. Björkstén, C. K. Lai, D. P. Strachan, S. K. Weiland, H. Williams, and the ISAAC Phase Three Study Group. 2006. "Worldwide Time Trends in the Prevalence of Symptoms of Asthma, Allergic

Rhinoconjunctivitis, and Eczema in Childhood: ISAAC Phases One and Three Repeat Multicountry Cross-Sectional Surveys." *Lancet* 368 (9537): 733–743.

Ayres, J. A. 2000. *Sensory Integration and the Child.* Los Angeles: Western Psychological Services.

BBC News. 2010. "Finland's Education Success." The World Videos 1. April 10. https://www.youtube.com/watch?v =rlYHWpRR4yc.

Biel, L., and N. K. Peske. 2009. *Raising a Sensory Smart Child: The Definitive Handbook for Helping Your Child with Sensory Processing Issues.* New York: Penguin.

Brody, J. E. 2009. "Babies Know: A Little Dirt Is Good for You." *New York Times*, January 26. http://www.nytimes.com/2009 /01/27/health/27brod.html.

Bundy, A. 1997. "Play and Playfulness. What to Look For." In *Play in Occupational Therapy for Children*, edited by L. D. Parham and L. S. Fazio. St. Louis, MO: Mosby.

Campbell, D. 2011. "Children Growing Weaker as Computers Replace Outdoor Activity." *Guardian*, May 21. http://www.theguardian.com/society/2011/may/21/children -weaker-computers-replace-activity.

Case-Smith, J. 2001. *Occupational Therapy for Children.* St. Louis: Mosby.

Centers for Disease Control and Prevention. 2010. *The Association Between School-Based Physical Activity, Including Physical Education, and Academic Performance.* Atlanta, GA: US Department of Health and Human Services.

Clements, R. 2004. "An Investigation of the Status of Outdoor Play." *Contemporary Issues in Early Childhood* 5 (1): 68–80.

Cohen, L. J. 2013. "The Drama of the Anxious Child." *Time,* September 26. http://ideas.time.com/2013/09/26/the-drama -of-the-anxious-child.

Crawford, N. 2013. "Katie Bowman and the Biomechanics of Human Growth: Barefoot Babies." *Breaking Muscle.com.* http://breakingmuscle.com/family-kids/katy-bowman -and-the-biomechanics-of-human-growth-barefoot-babies.

Dewar, G. 2011. "When 'Daycare' was Run by Kids." *Babycenter Blog.* March 4. http://blogs.babycenter.com/mom_stories /when-daycare-was-run-by-kids.

Dunn, R., and K. Dunn. 1993. *Teaching Secondary Students Through Their Individual Learning Styles: Practical Approaches for Grades 7–12.* Michigan: Allyn and Bacon.

Fearn, H. 2015. "Child Kidnappings and Abductions Could Be Four Times Higher than Authorities Admit, Charities Warn." *Independent.* February 21. http://www.independent. co.uk/news/uk/crime/child-kidnap-and-abduction-increase- as-crimes-come-under-greater-scrutiny-10062014.html.

Fisher, A. V., K. E. Godwin, and H. Seltman. 2014. "Visual Environment, Attention Allocation, and Learning in Young Children: When Too Much of a Good Thing May Be Bad." *Psychological Science* 25 (7): 1362–1370.

Frick, S. M., and S. R. Young. 2012. *Listening with the Whole Body: Clinical Concepts and Treatment Guidelines for Therapeutic Listening.* Madison, WI: Vital Links.

Geoffroy, M. C., S. M. Côté, S. Parent, and J. R. Séguin. 2006. "Daycare Attendance, Stress, and Mental Health." *Canadian Journal of Psychiatry* 51 (9): 607–615.

Ginsburg, K. R. 2007. "The Importance of Play in Promoting Healthy Child Development and Maintaining Strong Parent-Child Bonds." *Pediatrics* 119 (1): 182–191.

Grahn P., F. Martensson, B. Llindblad, P. Nilsson, and A. Ekman. 1997. "Ute på Dagis." *Stad and Land* 145. Håssleholm, Sweden: Nora Skåne Offset.

Gray, P. 2013. *Free to Learn: Why Unleashing the Instinct to Play Will Make Our Children Happier, More Self-Reliant, and Better Students for Life.* New York: Basic Books.

Greenfield, B. 2015. "Parents Under Investigation for Neglect." *Yahoo Parenting*, January 15. https://www.yahoo.com /parenting/parents-under-investigation-for-neglect-after -108180228512.html.

Hamilton, J. 2014. "Scientists Say Child's Play Helps Build a Better Brain." NPR Ed. August 6. http://www.npr.org /sections/ed/2014/08/06/336361277/scientists-say-childs -play-helps-build-a-better-brain.

Hanford, E. 2015. "Out of the Classroom and Into the Woods." NPR Ed. May 26. http://www.npr.org/sections/ed/2015/05/26 /407762253/out-of-the-classroom-and-into-the-woods.

Harris, E. A. 2015. "Sharp Rise in Occupational Therapy Cases at New York's Schools." *New York Times*, February 17. http://www.nytimes.com/2015/02/18/nyregion/new-york-city -schools-see-a-sharp-increase-in-occupational-therapy-cases .html.

Harris, L. No date. "Are Too Many Kids Receiving Occupational Therapy?" Babble.com. http://www.babble.com/kid/are-too-many-kids-receiving-occupational-therapy.

Harrison, K., T. Harrison, and K. McArdle. 2013. "Outdoor Play and Learning in Early Childhood from Different Cultural Perspectives." *Journal of Adventure Education and Outdoor Learning* 13 (3): 238–254.

Hedström, E. M., O. Svensson, U. Bergström, and P. Michno. 2010. "Epidemiology of Fractures in Children and Adolescents." *Acta Orthopaedica* 81 (1): 148–153.

Hubbard, S. B. 2005. "Are too Many Vaccines Destroying Kids' Immune Systems?" Newsmax.com. February 5. http://www.newsmax.com/Health/Headline/vaccines-children-immune-system/2015/02/05/id/622900.

Jarrett, O. S., D. M. Maxwell, C. Dickerson, P. Hoge, G. Davies, and A. Yetley. 1998. "Impact of Recess on Classroom Behavior: Group Effects and Individual Differences." *Journal of Educational Research* 92 (2): 121–126.

Jensen, E. 1998. *Teaching with the Brain in Mind.* Alexandria, VA: Association for Supervision and Curriculum Development.

Juster, F. T., H. Ono, and F. P. Stafford. 2004. *Changing Times of American Youth: 1981–2003.* Ann Arbor, MI: Institute for Social Research. http://ns.umich.edu/Releases/2004/Nov04/teen_time_report.pdf.

Kable, J. 2010. "Theory of Loose Parts." *Let the Children Play* (blog). February 10. http://www.letthechildrenplay.net/2010/01/how-children-use-outdoor-play-spaces.html.

Kawar, M. J., and S. M. Frick. 2005. *Astronaut Training: A Sound Activated Vestibular-Visual Protocol for Moving, Looking, and Listening.* Madison, WI: Vital Links.

Kelley, B., and C. Carchia. 2013. "Hey Data, Data—Swing!" ESPN.com. July 16. http://espn.go.com/espn/story/_/id /9469252/hidden-demographics-youth-sports-espn-magazine.

Kraft, R. E. 1989. "Children at Play: Behavior of Children at Recess." *Journal of Physical Education, Recreation, and Dance* 60 (4): 21–24.

Kranowitz, C. S. 1998. *The Out-of-Sync Child: Recognizing and Coping with Sensory Integration Dysfunction.* New York: Perigee Books.

Lee, H. 2013. "The Babies Who Nap in Sub-Zero Temperatures." *BBC News Magazine,* February 22. http://www.bbc.com/news/magazine-21537988.

Levy, A. 2013. "Parents' Anxieties Keep Children Playing Indoors: Fears About Traffic and Strangers Leading to 'Creeping Disappearance' of Youngsters from Parks." Daily Mail.com. August 6. http://www.dailymail.co.uk/news /article-2385722/Parents-anxieties-children-playing-indoors -Fears-traffic-strangers-leading-creeping-disappearance -youngsters-parks.html.

Manier, J. 2008. "Experts Say 'Tummy Time' Key for Tots." *Chicago Tribune,* January 27. http://articles.chicagotribune .com/2008–01–27/news/0801270067_1_tummy-time -benign-source-bouncy-seat.

National Space Biomedical Research Institute. No date. "The Body in Space." http://www.nsbri.org/DISCOVERIES-FOR -SPACE-and-EARTH/The-Body-in-Space.

Nationwide Children's. No date. "Dance-Related Injuries by the Numbers." http://www.nationwidechildrens.org/dance-injuries-by-the-numbers.

Nationwide Children's. 2009. "New National Study Finds Increase in P. E. Class-Related Injuries." August 3. http://www.nationwidechildrens.org/news-room-articles/new-national-study-finds-increase-in-pe-class-related-injuries?contentid=49229.

Neate, R. 2013. "Campfire Kids: Going Back to Nature with Forest Kindergartens." *Spiegel Online International*, November 22. http://www.spiegel.de/international/zeitgeist/forest-kindergartens-could-be-the-next-big-export-from-germany-a-935165.html.

Nussbaum, D. 2006. "Before Children Ask, 'What's Recess?'" The New York Times on the Web Learning Network. December 10. https://www.nytimes.com/learning/students/pop/20061214snapthursday.html.

Ogden, C. L., M. D. Carroll, B. K. Kit, and K. M. Flegal. 2012. *Prevalence of Obesity in the United States, 2009–2010.* National Center for Health Statistics Data Brief 82. US Department of Health and Human Services, Centers for Disease Control and Prevention. Retrieved from http://www.cdc.gov/nchs/data/databriefs/db82.htm.

Ohio State University College of Optometry. 2014. "Scientists Study Effects of Sunlight to Reduce Number of Nearsighted Kids." November 20. http://optometry.osu.edu/news/article.cfm?id=330.

Okada, H., C. Kuhn, H. Feillet, and J. F. Bach. 2010. "The 'Hygiene Hypothesis' for Autoimmune and Allergic

Diseases: An Update." *Clinical and Experimental Immunology* 160 (1): 1–9.

Palmer, B. 2013. "Why Are So Many Kids Getting Myopia?" *Slate*, October 16. http://www.slate.com/articles/health_and _science/medical_examiner/2013/10/myopia_increasing _indoor_light_may_be_impairing_children_s_vision.html.

PBS Parents. No date. "Raising a Powerful Girl." http://www.pbs.org/parents/parenting/ raising-girls/body -image-identity/raising-a-powerful-girl.

Pellegrini, A. D. 2009. "Research and Policy on Children's Play." *Child Development Perspectives* 3 (2): 131–136.

Pellegrini, A. D., and C. M. Bohn-Gettler. 2013. "The Benefits of Recess in Primary School." *Scholarpedia* 8 (2): 30448.

Pica, R. 2010. "Why Kids Need Recess." *Pathways to Family Wellness* 25. http://pathwaystofamilywellness.org/Children-s -Health-Wellness/why-kids-need-recess.html.

Rao, U. B., and B. Joseph. 1992. "The Influence of Footwear on the Prevalence of Flat Foot: A Survey of 2300 Children." *Journal of Bone and Joint Surgery* 74-B (July): 525–527.

Reddy, S. 2015. "The Benefits of Fidgeting for Students with ADHD." *Wall Street Journal*, June 22. http://www.wsj.com /articles/the-benefits-of-fidgeting-for-students-with-adhd -1434994365.

Reed, A. C., T. M. Centanni, M. S. Borland, C. J. Matney, C. T. Engineer, and M. P. Kilgard. 2014. "Behavioral and Neural Discrimination of Speech Sounds After Moderate or Intense Noise Exposure in Rats." *Ear and Hearing* 35 (6). Retrieved from http://www.researchgate.net/publication/264393720

_Behavioral_and_Neural_Discrimination_of_Speech _Sounds_After_Moderate_or_Intense_Noise_Exposure _in_Rats.

Rideout, V. J., U. G. Foehr, and D. F. Roberts. 2010. *Generation M2: Media in the Lives of 8- to 18-Year-Olds.* A Keiser Family Foundation Study. Retrieved from https://kaiserfamily foundation.files.wordpress.com/2013/04/8010.pdf.

Roley, S. S., E. I. Blanche, and R. C. Schaaf. 2001. *Understanding the Nature of Sensory Integration with Diverse Populations.* San Antonio, TX: Therapy Skill Builders.

Rosin, H. 2014. "The Overprotected Kid." *Atlantic,* April. http://www.theatlantic.com/magazine/archive/2014/04 /hey-parents-leave-those-kids-alone/358631.

Russia Today, 2015. "UK Child Kidnappings and Abductions Soar By 13%." February 22. https://www.rt.com /uk/234499-child-abductions-kidnappings-increase/.

Sandseter, E., L. E. O. Kennair. 2011. "Children's Risky Play from an Evolutionary Perspective: The Anti-Phobic Effects of Thrilling Experiences." *Evolutionary Psychology 9 (2):* 257–284. Retrieved from http://www.epjournal.net/ articles /children's-risky-play-from-an-evolutionary-perspective-the -anti-phobic-effects-of-thrilling-experiences/getpdf.php ?file=EP092572842.pdf.

Saul, H. 2014. "New Zealand School Bans Playground Rules and Sees Less Bullying and Vandalism." *Independent,* January 28. http://www.independent.co.uk/news/world/australasia/new -zealand-school-bans-playground-rules-and-sees-less-bullying -and-vandalism-9091186.html.

Scrivens, D. 2008. "Rebounding: Good for the Lymphatic System." *Well Being Journal* 17 (3). Retrieved from https://www.wellbeingjournal.com/rebounding-good-for-the-lymph-system.

Shaw, B., B. Watson, B. Frauendienst, A. Redecker, T. Jones, and M. Hillman. 2013. *Children's Independent Mobility: A Comparative Study in England and Germany (1971–2010).* London: Policy Studies Institute.

Skenazy, L. 2009. *Free-Range Kids: How to Raise Safe, Self-Reliant Children (Without Going Nuts with Worry).* San Francisco, CA: Jossey-Bass.

Southern Illinois University School of Medicine. 2007. "Weak Bones in Children." Press release, December 11. Retrieved from http://www.siumed.edu/news/Newsline%20TEXT08 /12–11–07.html.

St. George, D. 2015. "Parents Investigated for Neglect After Letting Kids Walk Home Alone." *Washington Post*, January 14. http://www.washingtonpost.com/local/education /maryland-couple-want-free-range-kids-but-not-all-do/2015 /01/14/d406c0be-9c0f-11e4-bcfb-059ec7a93ddc_story.html.

Stevenson, P. 2006. "Banning Tag at Recess Is Dumb." CBS News. June 27. http://www.cbsnews.com/news/banning-tag -at-recess-is-dumb.

Szabo, L. 2011. "One in Six Children Have a Developmental Disability." *USA Today*, May 22. http://usatoday30.usatoday .com/news/health/story/health/story/2011/05/One-in -six-children-have-a-developmental-disability/47467520/1.

Taylor, A. F., F. E. Kuo, and W. C. Sullivan. 2001. "Coping with ADD: The Surprising Connection to Green Play Settings." *Environment and Behavior* 33 (1): 54–77.

Thai Teachers Television. 2012. "Finland Lessons—Secondary Science, the Fishing Line." NeoEdutainment. January 22. https://www.youtube.com/watch?v=pkNEVNNeWdQ.

Theodore Roosevelt Association. No date. "Quotations." To Cuno H. Rudolph, Washington Playground Association (WPA), February 16, 1907. Presidential Addresses and State Papers VI, 1163. Retrieved from http://www.theodore roosevelt.org/site/c.elKSIdOWIiJ8H/b.9297493/k.37C4 /Quotations.htm.

Tierney, J. 2011. "Can a Playground Be Too Safe?" *New York Times*, July 18. http://www.nytimes.com/2011/07/19/science /19tierney.html?_r=0.

Tourula, M., A. Isola, and J. Hassi. 2008. "Children Sleeping Outdoors in Winter: Parents' Experiences of a Culturally Bound Childcare Practice." *International Journal of Circumpolar Health* 67 (2–3): 269–78.

Ulrich, R. S. 1984. "View Through a Window May Influence Recovery from Surgery." *Science, New Series* 224 (4647): 420–421. Retrieved from http://www.majorhospital foundation.org/pdfs/View%20Through%20a%20Window.pdf.

University of Maryland Medical Center. 2011. "Aromatherapy." August 9. http://umm.edu/health/medical/altmed/treatment/ aromatherapy.

US Food and Drug Administration. 2015. "Asthma: The Hygiene Hypothesis." January 26. http://www.fda.gov

/BiologicsBloodVaccines/ResourcesforYou/Consumers/ ucm167471.htm.

Visser, S. N., M. L. Danielson, R. H. Bitsko, J. R. Holbrook, M. D. Kogan, R. M. Ghandour, R. Perou, and S. J. Blumberg. 2013. "Trends in the Parent-Report of Health Care Provider-Diagnosed and Medicated Attention-Deficit/Hyperactivity Disorder: United States, 2003–2011." *Journal of the American Academy of Child and Adolescent Psychiatry* 53 (1): 34–46.

Wang, B. 2013. "More Schools Banning 'Tag' Because of Injuries." *FindLaw* (blog). October 11. http://blogs.findlaw .com/injured/2013/10/more-schools-banning-tag-because -of-injuries.html.

Zygmunt-Fillwalk, E., and T. E. Bilello. 2005. "Parents' Victory in Reclaiming Recess for Their Children." *Childhood Education* 82 (1): 19–23.

Angela J. Hanscom is a pediatric occupational therapist and founder of TimberNook—an award-winning developmental and nature-based program that has gained international popularity. She holds a master's degree in occupational therapy, and an undergraduate degree in kinesiology (the study of movement) with a concentration in health fitness. Awarded a "Hometown Hero" by *Glamour* magazine for her innovative work with TimberNook, Hanscom has also been a frequent contributor to *The Washington Post*'s "Answer Sheet" column, and was featured on the NPR education blogs *Children & Nature Network* and *MindShift*. Hanscom resides in Barrington, NH.

Foreword writer **Richard Louv** is a journalist, and author of *Last Child in the Woods* and *The Nature Principle*. He is cofounder of the Children and Nature Network—an organization helping to connect people to the natural world. He's written for *The New York Times*, *The Washington Post*, and more, and appears on programs such as NBC's *Today* and NPR's *Fresh Air*.

Index

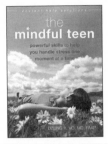
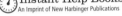